Industrial Control Electronics

Industrial Control Electronics

David P. Beach
Indiana State University

Roy G. Bridges
International Brotherhood of Electrical Workers
Local #725, Indiana

PRENTICE HALL, Englewood Cliffs, New Jersey 07632

Library of Congress Cataloging-in-Publication Data

BEACH, DAVID P., (Date)
Industrial control electronics / David P. Beach, Roy G. Bridges.
p. cm.
ISBN 0-13-459256-5
1. Electronic control. 2. Industrial electronics. I. Bridges,
Roy G. II. Title.
TK7881.2.B4 1990 89-23214
621.31'7—dc20 CIP

Editorial/production supervision
and interior design: *Carol L. Atkins*
Cover design: *Wanda Lubelska Design*
Manufacturing buyer: *Gina Chirco-Brennan*

A Division of Simon & Schuster
Englewood Cliffs, New Jersey 07632

Printed in the United States of America
10 9 8 7 6 5 4 3 2 1

ISBN 0-13-459256-5

Prentice-Hall International (UK) Limited, *London*
Prentice-Hall of Australia Pty. Limited, *Sydney*
Prentice-Hall Canada Inc., *Toronto*
Prentice-Hall Hispanoamericana, S.A., *Mexico*
Prentice-Hall of India Private Limited, *New Delhi*
Prentice-Hall of Japan, Inc., *Tokyo*
Simon & Schuster Asia Pte. Ltd., *Singapore*
Editora Prentice-Hall do Brasil, Ltda., *Rio de Janeiro*

With affection and appreciation
this book is dedicated
to our families

Emily Kim Romily
and
Susan Danielle Andrew

Contents

Preface

The concept of electrical controls is not new, but the availability of new devices and applications has grown. The basic operation of the simple on-off switch has been expanded to include the use of computers. In our experiences as both electricians and educators it has been observed that a shroud of mystery still seems to overshadow the simplicity of the basic electrical control circuit. Even complex electrical circuits can be reduced to the Ohm's Law equation. The same is true with control circuits. If the circuit is examined closely the ultimate control operation is either an open or closed device, generally a switch. With this in mind, our first objective is to remove the mystery from electrically controlled circuits and present them in simple, basic terms or concepts that require no higher level mathematics. A second goal is to develop complex circuits from uncomplicated examples that gradually build upon basic concepts and then to reapply these principles to more intricate circuits and still retain the basic ideas. The intent of the book is to allow *anyone with a desire to learn* to have the ability to function in a control environment with confidence and to have logical reasoning for applying the principles presented. The emphasis is again on keeping it simple and not making the information complicated. This is important to the new learner who is just starting a career in the industrial control environment and equally as useful to the veteran who wishes to review old principles and apply them to new applications.

Chapter 1 introduces typical control devices by illustrating them and explaining their operation. Terms and practical applications are also presented. This is helpful to the novice who is seeing these items for the first time. It also provides an excellent starting point for a training program by explaining why each device has been chosen for a particular application.

Chapters 2 and 3 get to the "heart of the matter" by presenting the different types of electrical diagrams that are necessary for working with control circuits. Electrical symboling and correct methods of using start-stop circuit controls are used as an initial step toward what is called "control thinking." This mental process helps the reader to make the transition from the electrical diagram to the actual wiring of the circuit, and then to the final process for which the circuit was designed by breaking the circuit into its controlling elements.

Since many control circuits are eventually modified, Chapter 4 allows the reader to gain some flexibility in circuit design by introducing changes in circuit parameters and also in meeting design criteria—the "What do we want the circuit to do?" part of the design process. The ability to design a circuit and then have it function properly is rewarding and good design reinforces the fact that control circuits are very useful.

Upgrading of a relay circuit to a solid state logic circuit is the purpose of Chapter 5. The hybrid relay and its applications as well as some design considerations are explored. Logic circuits are discussed with the use of truth tables to predict the operation of a sequence of control inputs to a circuit.

Chapter 6 culminates the use of relays and logic circuits with a practical application of each device discussed. The transition from high- and low-level voltage sources to relay and logic functions is shown with the use of converters.

Chapter 7 begins another transition from relay logic to programmable controllers. This section is devoted primarily to "What is the black box called a programmable controller?" Although basic elements of a controller are discussed, the intricate circuitry involved intentionally eliminated. This is part of the "keep it simple principle" on which the book is based.

Chapters 8 and 9 are devoted to beginning principles upon which all programming is based. Generic programs, that are not devoted to a specific manufactured model or brand of programmable controller, are developed to allow the learner, with minimal effort, to apply the principles presented to any one of the many products available. Conversion from ladder diagrams to programming diagrams is emphasized. The use of peripheral equipment such as "master-control relays" and their function is also discussed.

Important aspects of controller applications are developed in Chapter 10.

Items necessary to properly install a programmable controller and to provide a trouble-free environment are examined. Consideration is also given to equipment operating near the controller.

The process of using existing relay components and applying them as inputs to a PC is the topic of Chapter 11. The expansion of outputs to a central controller or processor is of increasing importance. The use of interfacing circuits and future trends likely to be encountered when upgrading the automation process are presented along with protocol standards that are in existence or in development.

Chapter 12 is for those who must purchase or specify what equipment to buy, as well as those who wish to upgrade their knowledge of equipment specifications. Sample design specifications are given and discussed.

The various power systems that may be encountered in the industrial environment are listed in Chapter 13. Transformer operation and simple calculations for the various systems are presented. Current and voltage calculations are also shown to compare and contrast the different systems.

Chapter 14 emphasizes the importance of proper short-circuit protection. This short, but thorough, chapter discusses fuse sizing and possible high currents that may be encountered from incorrect selections. Examples are given with illustrations. Generic fuse characteristic curves are also provided with examples of their usage.

Since motors are a common load in the industrial area, Chapter 15 has a brief discussion on the operation of the various types of motors that are likely to be encountered. Characteristics and safe operating procedures are additionally included. Illustrations also show the various voltage connections commonly found.

Chapter 16 returns to the operation of typical control devices, primarily the motor starter, and some common applications. Different voltage starting methods are presented as well as the selection process necessary for proper sizing of electrical motor starters.

Chapter 17 has an in-depth presentation of relays. Typical uses and expectations that could be encountered are illustrated. Again proper selection techniques and usage are given consideration.

Chapter 18 expands on the simple relay circuits developed earlier in the book and discusses the use of speed controls in the motor circuit. Various types of speed-control circuits are illustrated.

Chapter 19 serves as an introduction to the control system as a theoretical process. The uses of feedback loops are shown as well as methods of controlling a process along with the uses of simple on-off controls. Primarily introductory in nature, it helps to illustrate that complex applications are possible with control elements.

Chapter 20 is an introduction to robotics. What a robotic unit is, how it is used, and what future applications are anticipated, are several factors discussed.

Not to be excluded is the need for safe operating procedures when using control circuits. There is emphasis throughout the entire text on safe operation of all electrical circuits.

The reader will find the text easy to read, and it will help to make the learning process enjoyable, as well as worthwhile.

David P. Beach
Roy G. Bridges

1

Control Devices

OBJECTIVES

Upon successful completion of this chapter, you should be able to

(1) describe the operation of an on-off switching device
(2) specify requirements for the selection of a particular switch for an application
(3) select an available switch and apply it to an industrial situation
(4) describe the operation of a relay
(5) use a relay in an industrial application
(6) describe the operation of a motor starter and contactors
(7) troubleshoot and service various kinds of switches, relays, actuators, and motor starters
(8) distinguish between normally-closed and normally-opened switches
(9) demonstrate comprehension of the safety rules and regulations that should be observed when using switches, contactors, and motor starters

As you begin your study of control circuits, many new terms and descriptions will confront you. Do not be overwhelmed by this. We have tried throughout the text to present items in simple terms, where possible, and then build upon them to present more complex subjects. The key to your success will be to study each section and review it whenever you feel your confidence lacking.

INTRODUCTION

The ON/OFF control of electronic or electrical circuits is a vital component for all stages of industrial electronics. Components used for such control are usually grouped into three major categories: switches, relays, and circuit breakers. Even though these devices are electromechanical systems, they have major applications in electronics. The "real electronic switching" that uses solid-state devices (such as transistors and integrated circuits) comprises another major category and will be discussed in later chapters.

The above grouping is not comprehensive nor exclusive—exclusive in the sense that items in each group may not be given the name of another item in another group. For example, a multiple-contact switching device shown in some catalogs as a stepping relay is shown in other catalogs as a stepping switch. Similarly, a reed switch that is actuated by a magnet or an energized coil can be found listed as a switch or as a relay.

Switches and relays are alike in their *on-off* or making and breaking actions. Circuit breakers are special normally-closed switches which are automatically opened as a result of electrical overload and can be manually reset. There are a number of types and forms of switches used for on-off control of electric circuits. They may be grouped according to mode of operation, construction, or by function.

Let's begin by dividing the control circuit into two-separate sections. The part of the circuit that is being controlled is called the final *control element*. This might be a relay, light, timer, or any other device that requires voltage to operate it. The other part of the circuit is the *control circuit*. All components used in the control circuit are usually called pilot or control devices. Generally these are either on, off, opened or closed. To provide greater flexibility for control circuits, time is varied by mechanical or electronic means. For example, a toggle switch is a pilot device.

For assisting you to understand some of the control systems that are available, we will discuss many of the more common devices. We shall start with pilot devices.

TOGGLE SWITCHES

The toggle switch is a very important pilot device. It provides us with the means of interrupting an electrical circuit without removing wires each time we wish to cycle something on or off. Toggle switches are used for manual control and as safety control devices. The simplest toggle switch has a handle and a single set of electrical contacts (the metallic points through which current flows) which open or close depending upon the handle position. More complicated versions can have many sets of contacts. Figure 1-1 illustrates a toggle switch and its electrical symbol.

Typical uses in the home are light switches which may be single pole (SPST) or three-way switches (SPDT). In the industrial environment they are used as ON/OFF controls or as selection controls. The abbreviations SPST and SPDT are very important when specifying a type of switch, since supply rooms or supply wholesalers may have hundreds of variations in stock. Even with the correct nomenclature, styles and shapes will vary considerably. Figure 1-2 illustrates some common variations. The intended use for the switch will often be a good guide for determining what style to choose. The type and magnitude of electrical current is another factor. Switches have to be able to withstand the current they must break. Alternating current (AC) and direct current (DC) ratings are not the same for a given switch. The type of load is also very important. Generally, loads are classified as inductive or noninductive. A motor is one example of an inductive load. Still another consideration is the location where the switch is to be mounted.

Switch Abbreviations

It was just stated that it is important to specify the type of switching or contact arrangement by using the appropriate abbreviation. The most common are listed in Table 1-1 with their abbreviations.

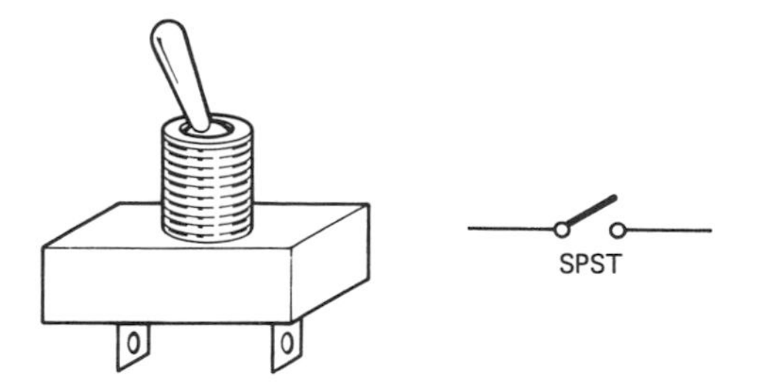

Figure 1-1 Toggle switch and symbol.

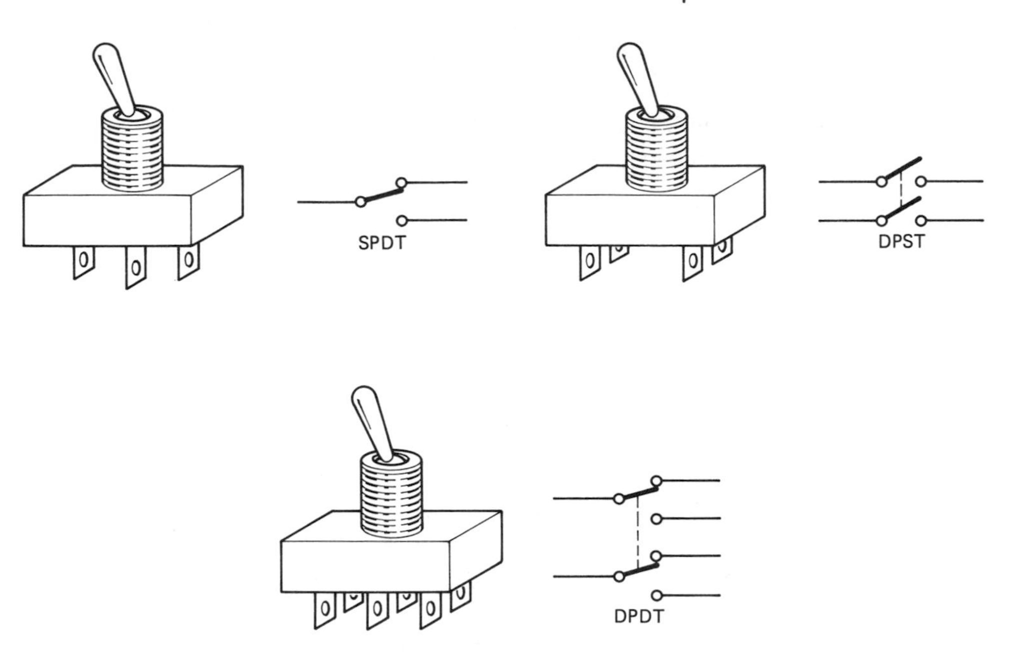

Figure 1-2 Toggle switch variations.

The term *pole* indicates the physical part that actually moves to make (close) or break (open) a circuit within the switch enclosure. The terms before the word *throw* indicate the number of possible positions, other than OFF, where the switch may be located or moved to complete circuit connections. A SPST switch would have only one movable part and only one possible point that breaks the circuit. A DPDT switch would have two movable parts with each one capable of touching two separate points.

In addition to the abbreviations, it should also be specified if the switch has momentary contacts or maintained contacts. This usually corresponds to the movement of the switch handle. A *maintained-contact switch* will remain in a position after the handle is released, while a *momentary-contact switch* will go back to a home position after it is released. A spring-return movement inside the

TABLE 1-1 SWITCH NOMENCLATURE

Terms	Abbreviations
Single Pole Single Throw	SPST
Single Pole Double Throw	SPDT
Double Pole Single Throw	DPST
Double Pole Double Throw	DPDT

switch causes it to go back to this position. This spring-return position is generally the OFF position, and the switch is often referred to as an *off return switch*.

Switch Mounting

The way the switch is mounted is important, especially if a defective switch is being replaced. The same mounting scheme will be used in most cases. When replacing a defective switch, remove the switch and take it with you while you look for a replacement. Having the old switch in hand can answer a lot of questions for you. It is good practice to replace equipment using the same mounting system in order to keep the orderly appearance of a control panel. A large variety of different equipment can soon make a control panel look unorganized and unprofessional.

Current Ratings

The current rating of the switch should match the current it will be interrupting. A switch rated at 15 amperes and 125 volts would not last very long in a circuit with a rating of 30 amperes. At a given voltage, The DC current rating of a switch would have a lower magnitude than the AC rating. Switch ratings are usually stamped or printed on the switch body. When in doubt look for the switch in a parts catalog and review the specifications given.

Inductive Loads

Inductive loads need higher rated current capabilities. The inrush or starting current may be four to six times the normal operating current. This could quickly damage a lower-rated switch. If the switch body has a label saying "noninductive loads only," do not use it on motor or other inductive loads. Some switches are also rated in *horsepower* and *voltage* to indicate that they are suitable for similar motor loads.

Switch Location

Switch location and environment are also important. Locations that are wet, have excessive dust, dirt, oil or explosive atmospheres need dust covers or switch bodies and mechanisms designed for these areas. These special considerations add to the initial cost of the switch but may prevent needless repairs later.

Other Considerations

Many other considerations are sometimes important. Some switches have illuminated handles to indicate the ON or OFF position. Life cycle ratings, types of terminals, vibration resistant contacts, contact resistance, temperature ratings and make-before-break switching are all factors that may need to be examined before selecting a switch. Some of these features might not be readily available with off-the-shelf items. If they are required and are considered critical items, they should be ordered in advance to assure that they will be available when needed. Total costs should include downtime of the process as well as failure rates of individual items.

SELECTOR SWITCHES

Selector switches have the same electrical characteristics as toggle switches. They differ in that to change the position of the switch the handle must be turned to the right or left. A typical selector switch is shown in Figure 1-3. Some selector switches are made with changeable mechanisms that allow different contact positions and additional mechanism bodies to be added in order for the same switch to have simultaneous changes in many different contact combinations.

An important feature of this type of switch is that it performs tasks that

Figure 1-3 Typical selector switches. (Courtesy of Allen-Bradley.)

would be extremely difficult with other types of switches: the simultaneous and sequential switching of many sets of contacts with the operation of a single actuator. Applications for this type of switch are in television tuners, multimeters, oscilloscopes, and other equipment. Simpler forms of selectable switches are used as step controls for heaters, lights, motors, and some industrial equipment.

PUSH-BUTTON SWITCHES

Push-button switches are available in momentary and maintained action. A combination provides a sequential action. A pull-to-operate switch, such as that used for automobile headlights, is a modification of the maintained-action type. Sequential action is obtained by cascading switches together in an assembly so that pressure on a push button actuates a switch and releases a previously actuated one.

Push-button switches, as the name implies, are pushed in order to change contact positions. They also have the same electrical characteristics as toggle switches. A variety of colors and styles are available. The most common use is for starting and stopping motors as illustrated in Figure 1-4. Stop buttons (push buttons) are usually red in color; many are raised or have large mushroom-type surfaces to allow easy access in emergencies. Start buttons are usually green to indicate a ''go'' condition. Illuminated buttons are also available to indicate which push-button function is in operation. In addition to momentary- and maintained-contact positions, some push buttons have a push-on and push-off feature. This is another version of the maintained-contact push button.

Figure 1-4 Start-stop push buttons. (Courtesy of Allen-Bradley.)

DRUM SWITCHES

Drum switches operate in a manner similar to toggle switches except that the operating mechanism is a "cam-like" device, and there are usually more than one set of contacts stacked around the cam. A handle is mounted on the cam to position it either to the right or left. Multiple positions are possible depending on the structure of the mechanism. Typical uses include starting and stopping of motors that also change direction of rotation and providing interlock circuits. Figure 1-5 shows a typical drum switch.

MICROSWITCHES

Microswitches are snap-acting switches in very small enclosures (see Figure 1-6). These small switches have the same electrical characteristics as toggle switches. The small size and variety of operating levers make them very useful as limit

(a) (b)

Figure 1-5 Drum switch. (a) External view (Courtesy of Square D Company). (b) Internal view (courtesy of Allen-Bradley).

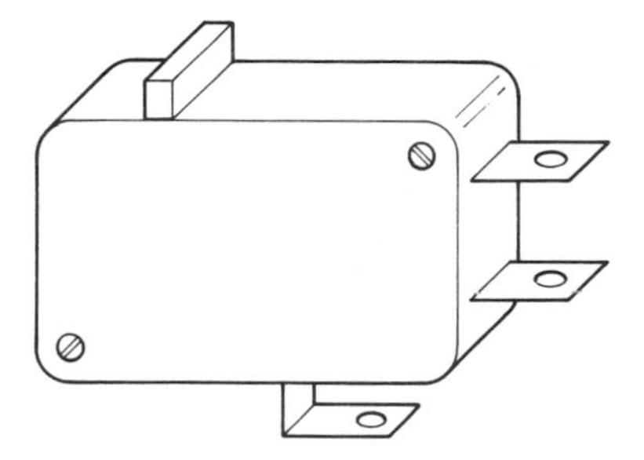

Figure 1-6 Microswitch.

switches. They are capable of operating with extremely small pressures on the operating levers, which allows a great deal of sensitivity. With the proper lever arrangement, even smaller pressures can operate the switch. Some switches have levers with rollers, flexible arms, and direct-contact actuators.

LIMIT SWITCHES

Limit switches are basically microswitches with heavy-duty cases for environmental protection. They generally define a function more than describe a type of switch. Any switch can be positioned to operate as a limiting device. Figure 1-7 shows a limit switch in an industrial type case.

Figure 1-7 Industrial limit switch. (Courtesy of Square D Company).

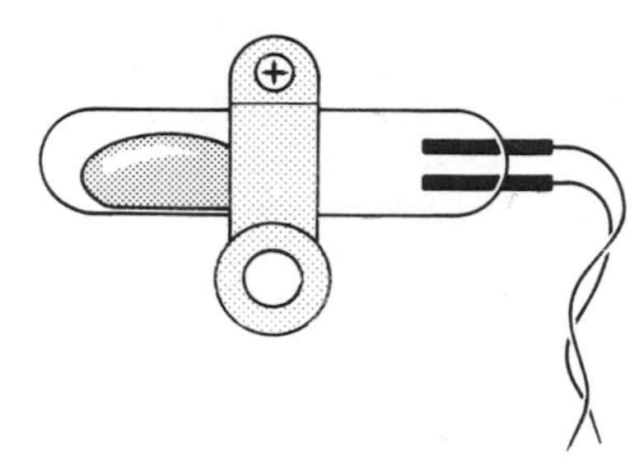

Figure 1-8 Mercury switch.

MERCURY SWITCHES

Mercury switches are very common in pressure-measuring devices. They consist of a glass bulb filled partially with mercury that covers two or more electrodes to *make* or *break* the circuit, as shown in Figure 1-8. By tilting the bulb, the mercury is allowed to move away from the contacts and break the circuit. A mounting mechanism that can be adjusted allows for precise operation at the appropriate angle. Your home thermostat is an example of its use. Multiple switches may be enclosed within the same bulb.

PRESSURE AND TEMPERATURE SWITCHES

Pressure and temperature switches are again microswitches adapted to pressure- or temperature-sensitive devices. Mercury switches are sometimes used for the same purpose.

PROXIMITY SWITCHES

Proximity switches are magnetically-sensitive devices used to operate a switch mechanism. A magnetic substance comes in close *proximity* to an encapsulated metal pole that bends to *make* or *break* a set of electrical contacts. They are used when no physical contact should be made with the item detected.

PHOTOCELL SWITCHES

Photocell switches use light-sensitive circuits to activate the switch mechanism. The light may be visible or infrared. Light is reflected or beamed to a receiver unit that senses its presence and then electronically activates a relay. Photocells can be made to function in either the presence or absence of a light beam. Figure 1-9

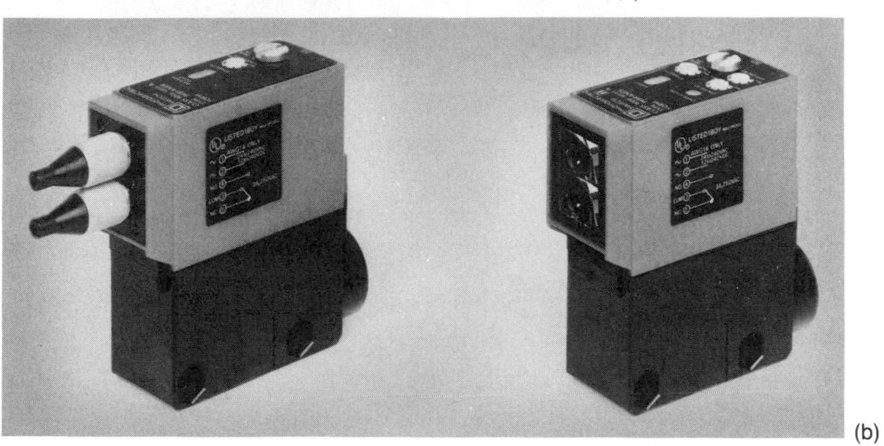

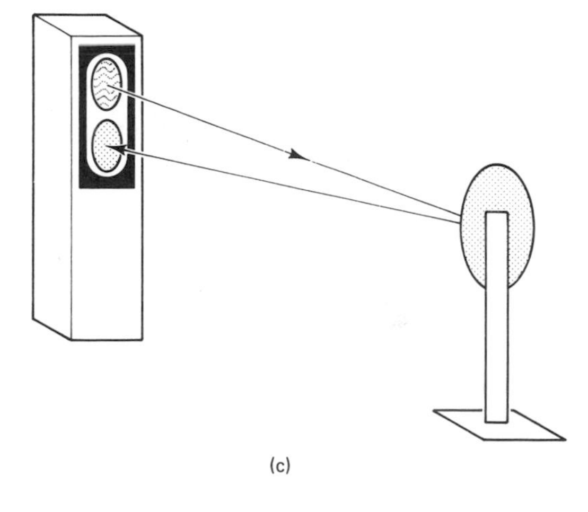

Figure 1-9 (a) Photocell transmitter and receiver (courtesy of Allen-Bradley); (b) Photocell transmitter and receiver (courtesy of Square D Company); (c) Photocell unit.

shows a common arrangement for the transmitter-receiver and reflector of a photocell.

FLOW SWITCHES

Flow switches are used to indicate the flow of a liquid. Fire-alarm sprinkler and electric pump systems are common uses. A paddle, in the liquid stream, bends with the force of the liquid flowing against it. Figure 1-10 illustrates a flow switch mounted in a pipe to indicate flow.

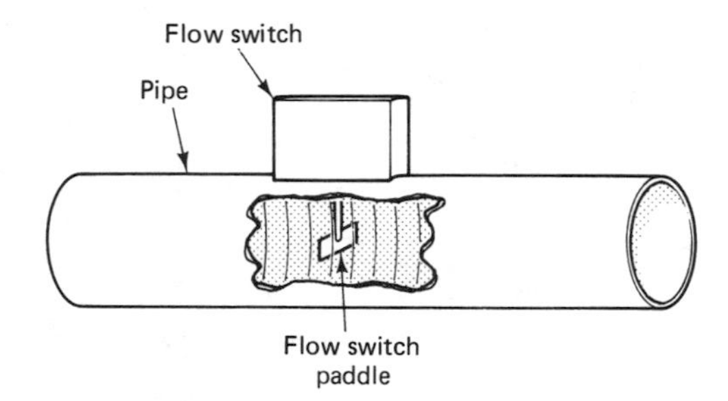

Figure 1-10 Flow switch.

DISCONNECTS

Switching low-power, low-current loads is not much of a problem, but as the current and power of a load increase, there soon becomes a practical physical limit for the size of the switch. Small 15- and 20-A loads can be handled by switches similar to those found in lighting circuits in the home. Once the current level passes the 20-A range, the size of the switch begins to increase at a fast rate. This increase in size is due to the physical size of the metal required to make the switch withstand higher levels of current. Switches in this category are called *disconnects*. Figure 1-11 illustrates one version of a disconnect switch used for small-current loads. Technically, any switch can be a "disconnecting device," but the term is usually reserved for the larger switches that must disconnect large current loads.

There are two types of disconnects, fused and nonfused. The fused disconnect has the added protection of a fuse to limit the amount of current flow in the circuit. This type of disconnect is often used as a *main disconnect* for large (high-current) circuits or distribution centers and as an isolating protection point in circuits. In automatic circuits, when maintenance must be performed at a remote point, a disconnect is used to *isolate* the device and protect the worker from

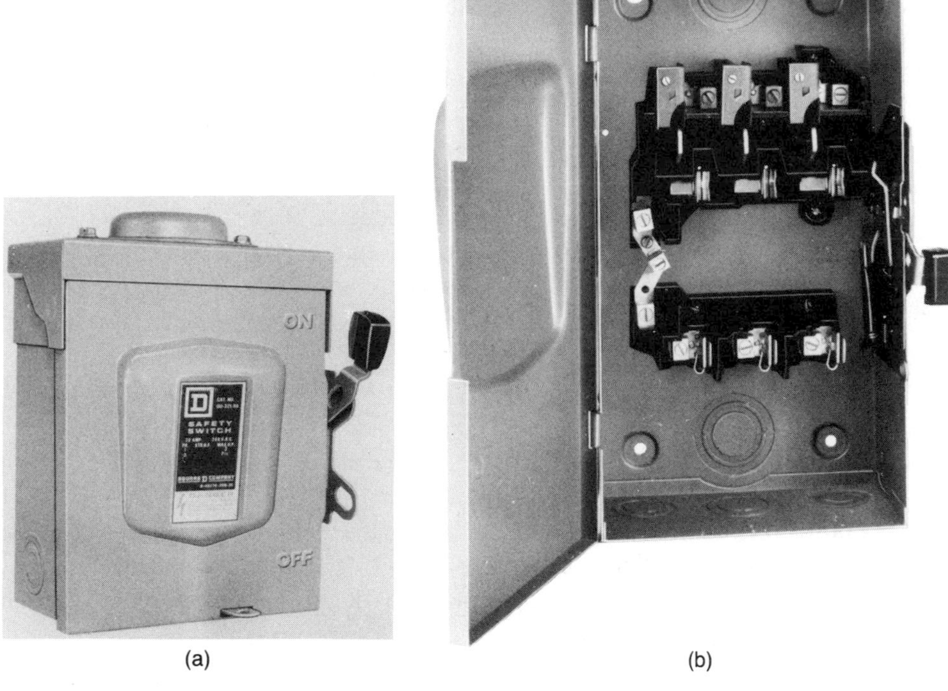

(a) (b)

Figure 1-11 (a) External view of a disconnect switch; (b) Internal view of a disconnect switch. (courtesy of Square D Company).

electrical shock. Disconnects used for this purpose do not have to be fused, but it may be convenient for the worker to be able to replace a fuse while at the machine instead of walking a great distance to replace it. Many of the electrical characteristics are the same as for a toggle switch. The values are just increased to reflect the need for handling larger currents and voltages. The most significant difference is the size of the switch. Instead of mounting the switch within a cabinet, the disconnect has its own enclosure which can be mounted as a separate unit. One final feature of the disconnect is that it can normally be locked in the open or closed position to prevent unauthorized use or as a safety precaution.

RELAYS

Even though there are a variety of switches available, they do have a limit to their usefulness. One such limit is that they must be physically moved to operate. This can be inconvenient at times, and a large area with nothing but disconnects is not only unattractive in appearance but takes up valuable floor and wall space that could be used for production purposes. Another device that can be used to perform a switching function is the relay. The relay is in all respects a switch, except that it is electrically operated instead of manually operated.

Relays are smaller in size than a comparable disconnect switch of the same rating. This enables many more relays to be placed within a given area. Since relays are electrically operated, they may be placed at remote locations. By using a control circuit, machines and devices in hazardous areas may be operated from points outside their dangerous environment. A worker would not have to go inside a radioactive container to turn on a motor, for example.

Relays have contacts which are parallel points for the current to flow through. These contacts may be round or square and made of silver, gold, or some alloy. Another part of the relay is the coil. When current flows through the coil, the metal pole inside is forced to move, and this in turn pushes or pulls one set of contacts against or away from another set of contacts to open or close a circuit. The electrical characteristics are the same as switches except that they also have a coil-voltage rating. Relay manufacturers provide several different coil-voltage ratings to allow more flexibility for the users. Since not all customers have the same voltage systems in their plants, one relay mechanism can be built and interchangeable coils can be wound to fit the various voltages used in control circuits.

Open and Enclosed Relays

Two common classifications are the *open* and *enclosed* types. Figure 1-12 shows both an open and enclosed type relay. The *open type relay* does not have any enclosure as part of the relay unit itself. The contacts, coil, and all moving parts are exposed to the open air and can be readily seen. They usually require some extra mounting hardware. This might be drilling and tapping holes for the mounting screws or drilling a hole for the mounting bolt of the relay. Most

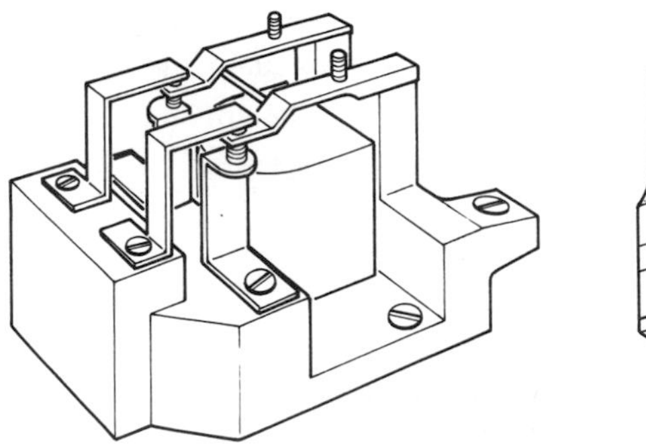

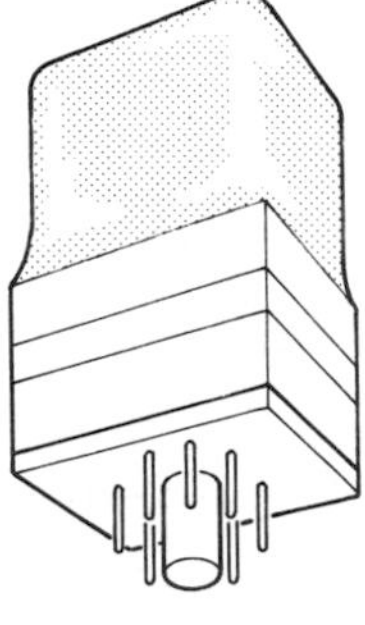

Figure 1-12 Open and enclosed relays.

mechanical relays are of the open type. A second type is the enclosed relay. *Enclosed relays* are either encapsulated or surrounded by a plastic cover. Encapsulated relays have an epoxy coating and generally are solder mounted to a printed circuit board. The plastic-covered relays have an outer shell of clear or opaque plastic and have a base (similar to a TV tube, on some units) which mounts in a socket. These are easily changed when needed but have small current-handling capabilities.

Reed Relays

The reed relay is another type of relay that is commonly used. These are found at the printed-circuit-board level. They switch low magnitude currents and voltages. It is their small size that allows them greater sensitivity than most relays. A glass-type reed relay is shown in Figure 1-13. These are not often found in older control systems; but if a printed circuit board can be found in a control circuit, it is very likely that a reed relay will be part of the circuit. A significant point is that, because of its small size and sensitivity, any large shock or vibration may cause the reed relay to open or close inadvertently. This should be considered when placing any printed circuit board in an industrial environment. The inadvertent operation of the relay could cause intermittent and premature cycling of other devices that are controlled by the relay. The reed relay is encapsulated to protect it from environmental conditions—no moving parts can be seen or heard.

Plug-In or Socket Relays

Although the *plug-in* or *socket* relay is small, it is much larger than the reed relay. It is used in control circuits that switch low-level current loads. Their chief advantage is the ease with which they may be changed if they become defective. Perhaps equally important is that the contacts are not exposed to corrosive environments. A mounting socket can be used either for a single relay or for a bank (line) of relays. Many combinations of contacts and voltages are available. Some plug-in relays are also available with time-delay functions. The *enclosed* relay in Figure 1-12 is a plug-in relay.

Time-Delay Functions

In some manufacturing processes it is necessary to have delays or time sequencing of the control cycle. This can be done with time-delay relays. Relays are

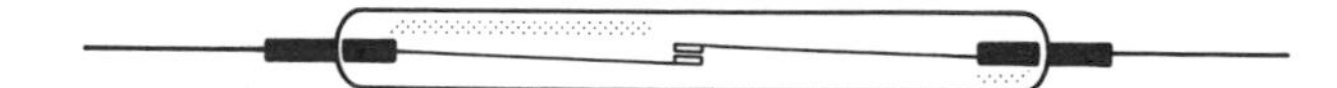

Figure 1-13 Glass-type reed relay.

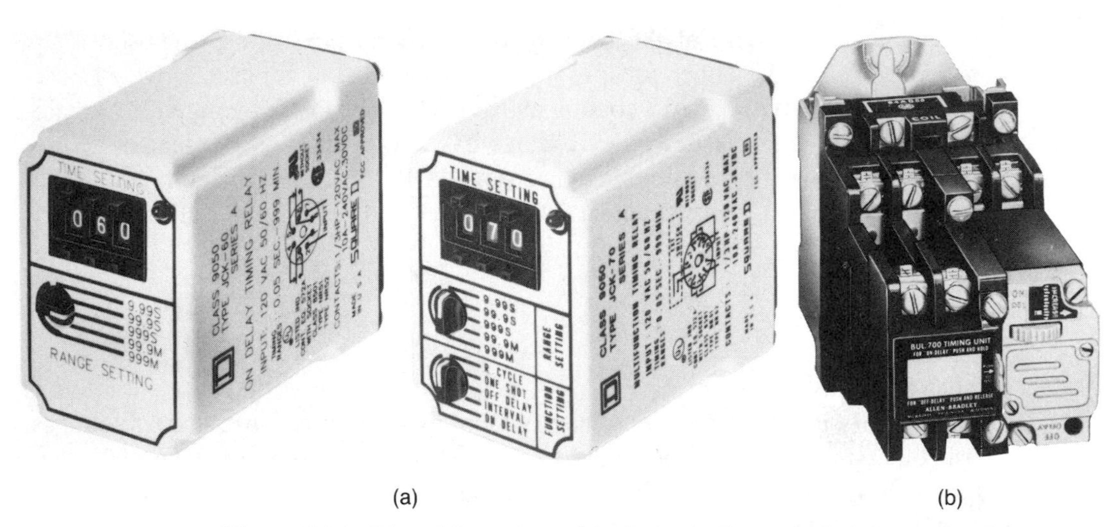

(a) (b)

Figure 1-14 Time-delay relays; (a) electronically controlled (courtesy of Square D Company), (b) Pneumatically controlled (courtesy of Allen-Bradley).

classified as time delay on energizing (TDOE) or time delay on de-energizing (TDOD). Some units have both capabilities. The time delay is created by allowing a small amount of air or liquid to pass through an orifice. Since a specified volume of air or liquid is required to operate the mechanism, changing the rate at which it leaves a reservoir also changes the time required for it to operate. Figure 1-14 illustrates an air-operated time-delay relay. Electronic circuits in solid-state relays enable them to perform the same functions with the advantage of no moving parts.

Contactors

For relays, just as with switches, when the current-handling capability increases so does the size of the device. Relays that switch large amounts of current are usually called *contactors*. Although there is a fine line between the point when a relay becomes a contactor, a good guide might be that if the current rating is greater than 40 A, it is probably a contactor. Contactors are physically larger and have more of a heavy-duty mechanical structure. To handle the larger currents, the contacts are much thicker and larger. A contactor is illustrated in Figure 1-15. Except for the current ratings, their function is the same as that of a relay. Typical uses include motor operation, switching large sections of lights from a single switch, and energizing distribution panels that contain night-lighting circuits.

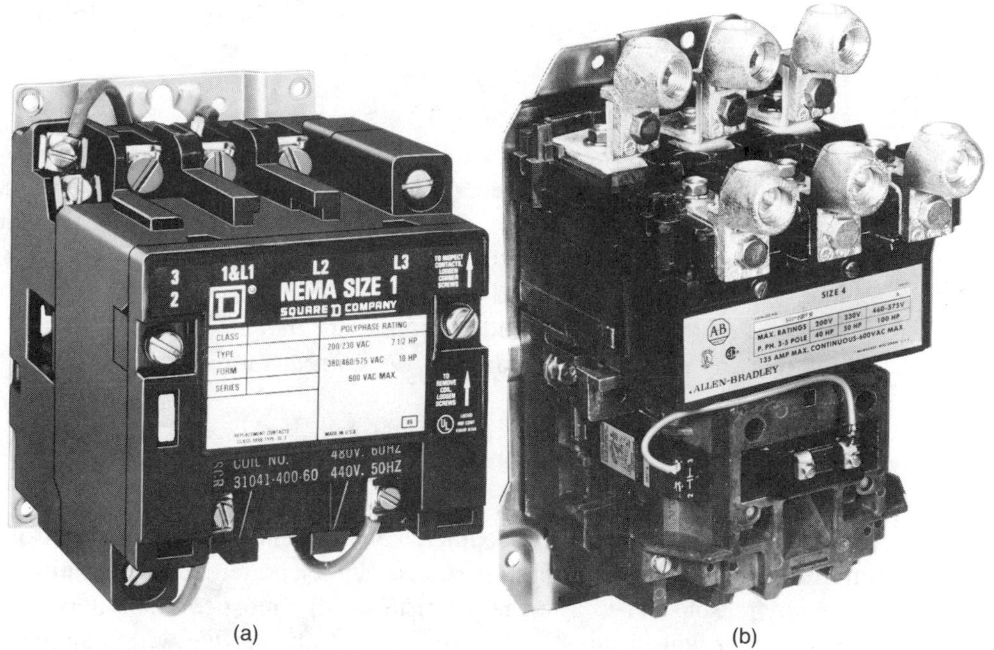

(a) (b)

Figure 1-15 Contactors. (a) Courtesy of Square D Company. (b) Courtesy of Allen-Bradley.

(a) (b)

Figure 1-16 (a) Motor starters assembled in a motor-control center (photo courtesy of Furnas Electric Company); (b) Motor starter (courtesy of Square D Company).

MOTOR STARTERS

Motor starters are relays or contactors with the added feature of overload protection included as part of the unit (see Figure 1-16). Almost all motors in an industrial environment will have a motor starter to start, stop and provide overload protection for it. The overload protection device mechanically releases the contact-holding mechanism and opens the electrical circuit when an overload occurs. Heat producing coils called *heaters* are mounted on the overload part of the starter and are sized according to the current rating of the motor.

SUMMARY

The type of control device encountered varies with the process controlled, the environment, availability of the part, and factory preference, to mention a few of the many reasons. The basic characteristics of the most common items have been discussed for you in this chapter. In the next chapter we will examine ways of representing these items in diagrams and some typical circuits that are found in an industrial environment.

EXERCISES

On a separate sheet of paper, complete a response for each question, statement, or problem listed below.

1.1. What name is given to the part of the circuit that is being controlled?

1.2. What is a pilot device? Provide an example.

1.3. Describe two major uses for a toggle switch.

1.4. List three factors that determine the type and style of toggle switch which should be used for a particular application.

1.5. Identify an example of an inductive load and an example of a noninductive load.

1.6. What does the term *pole* mean when referring to toggle switches, such as SPST?

1.7. Why is the current rating of a switch very important when choosing a switch for a particular application?

1.8. What could happen if you used a switch labeled noninductive loads only to switch an inductive load?

1.9. Why must adequate consideration be given to the switch location and environment?

1.10. Push-button switches are commonly used in ______ and ______ motors.

1.11. What is the significance of using *red* color for stopping and *green* for starting in most push-button applications?

1.12. Multiple positions are possible in some drum switches depending on the ______ of the mechanism.

1.13. What are the features that made the microswitches very useful?

1.14. Why are heavy-duty cases used on limit switches?

1.15. Describe an example of a mercury switch application.

1.16. What is a photocell switch?

1.17. What are the common applications for the flow switch?

1.18. Disconnect switches are more commonly used in switching ______ current loads.

1.19. The ______ is the most significance difference between a *disconnect* switch and other types of switches.

1.20. What is a relay?

1.21. Name two common classifications of the relay.

1.22. What is the major disadvantage of a reed relay?

1.23. What is the chief advantage of a plug-in or socket relay?

1.24. Identify what TDOE and TDOD stand for.

1.25. What are the main features of a contactor that allow it to handle large amounts of current?

1.26. What is the major advantage of a motor starter?

2

Control Diagrams

OBJECTIVES

Upon successful completion of this chapter, you should be able to

(1) identify and prepare common electrical control symbols
(2) recognize, describe, and prepare circuit illustrations that include associated symbols for indicating lamps, coils, solenoids, switches, relays, and conductors
(3) describe procedures for preparing control diagrams that include basic drafting standards and representations for two-wire and/or three-wire control systems

ELECTRICAL DIAGRAMS

Electrical diagrams are used to develop properly a common reference point for the transfer of circuit functions, relationships to other circuits, and normal circuit operation. Electrical diagrams are usually called electrical schematics or schematics. There are several different types of schematic diagrams. The most common type illustrates all of the electrical components and their relationship to other parts of the circuit. This type of drawing is used as an aid to locate a part number, a component value, or trace circuit wiring. The amount and type of information available on the drawing is related to functions of the end user. End user functions may be for basic information (functional diagrams), ordering repair parts, making interconnections to other equipment, or installing and repairing equipment.

Functional diagrams, sometimes called block diagrams, stress only the minimal information necessary or required to describe a process. Unlike the schematic, which shows all the components, the functional diagram represents the entire function or output of a single circuit as a block, with the appropriate label. This type of diagram is most useful in locating individual parts of a system, such as printed circuit boards, that are replaced as units. If a signal is going into a specific unit and no signal is coming out, there is a good possibility that the unit is defective (this example assumes a circuit with one input and one output, since more complex circuits require multiple signals as inputs and outputs). See Figure 2-1.

Once a defective part is located it is either repaired or replaced with a new part. Schematics label the part with *descriptive notation* to indicate what it does in the circuit, *value notation* to indicate the amount of a characteristic, and sometimes a part number for the component. Any one or all may be used to describe the part to the degree necessary to order a replacement. Figure 2-2 illustrates the amount of information available from each. Figure 2-2(a) shows a resistor with descriptive notation. The only information available is that the component is a resistor and someone chose to call it "R 20." Figure 2-2(b) shows the same resistor with value notation. With this information a replacement could easily be ordered. Figure 2-2(c) illustrates a resistor, but, to order it, the information derived must be referenced back to the firm designating the part number. Figure 2-2(d) shows a combination of descriptive notation and value notation. This is the most common method, and it provides the most information.

When large equipment is shipped for installation, it is often divided into several smaller units. To make the unit functional again, it must be interconnected as a single unit. To aid in the interconnecting process, schematics called wiring diagrams are used. When a great amount of detail is necessary for connecting the separate units, the workers need clear and concise directions; or when the

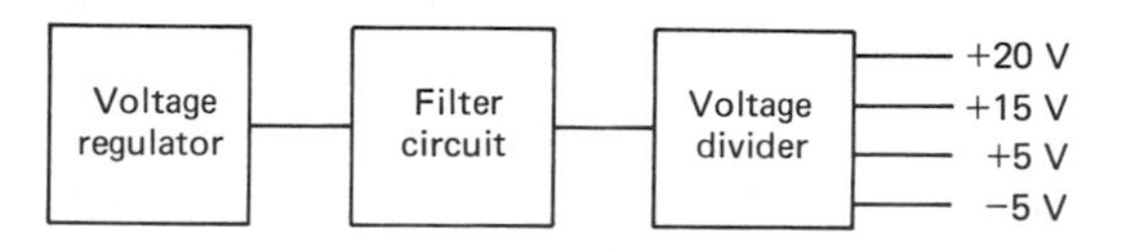

Figure 2-1 Functional diagram.

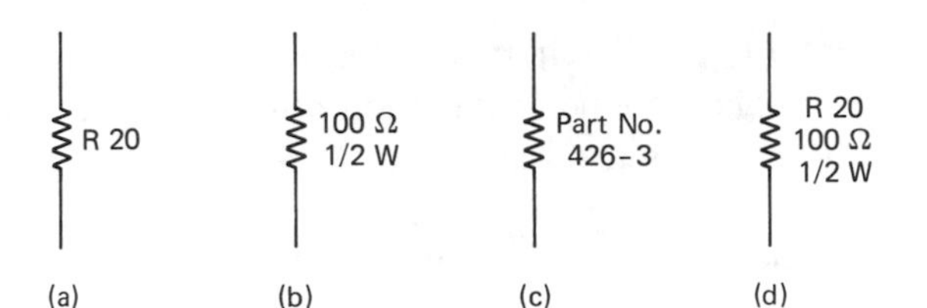

Figure 2-2 Schematic labels.

experience level requires extra guidance, the wiring diagram is most useful. Two types are the electrical symbol format and the pictorial format. The electrical symbol format uses standard electrical symbols and lines to indicate where wires are to be connected. The pictorial format may be a photograph or pictorial drawing. The electrical symbol format is the most commonly used method.

In an industrial environment many different circuits are used in the different machines. When a breakdown occurs, the schematic diagram can be of great value. By analyzing the problem that has occurred and relating it to the schematic diagram, the device causing the problem can usually be isolated. Since each machine may have its own unique process, the electrical diagram becomes even more important.

Many variations of schematics can be used and often the different versions are combined. The type of schematic found depends not only on who will use it but many times may simply be a matter of preference of the designer or engineer. Large firms tend to standardize their drawings.

Once a worker has experience and basic knowledge of wiring devices, it is not necessary to show all wiring on a drawing. The function of the circuit and the minimal wiring necessary can be illustrated on a *control diagram*. The only wiring found on a control diagram is that necessary to perform the control function of the device. An individual with a working knowledge of control diagrams can understand the process of a machine from the diagram alone. Since these diagrams are of importance in control environments, they will be discussed at some length.

Control Symbols

Before control diagrams can be examined, it is necessary to mention briefly control symbols. Several electronic symbols are used with control diagrams, but a symbol standard has also been developed specifically for the control environment. A list of common symbols can be found in the Appendix. The most common devices represented are switches, relay contacts, indicating lamps and relay coils or solenoids. Other symbols used are transformers, resistors, and fuses. These are not all of the symbols, but they are the most common.

Switches and Relay Contacts

All switches and relay contacts can be classified as normally open or normally closed. The positions drawn are the electrical characteristics of each device as they would be found when purchased and not connected in any circuit. This is often referred to as the *off-the shelf* position. It is important to understand this because it may also represent the *de-energized* position in a circuit. The de-energized position refers to the component position when the circuit is de-energized or no power is present on the circuit. This point of reference makes it possible to analyze new circuits when they are encountered.

The symbols used to represent the various functions have been derived from the devices that usually activate them. The float switch is used to indicate switching action caused by change in liquid level. A round float-like ball is attached to the contact arm of the switch in the symbol. An air pressure switch uses a cup-like symbol to signify catching the wind. Temperature-sensitive switches have a symbol that represents a heating element. Flow indicating switches have a wedge-shaped symbol. Switches with a time-delay function have arrows. An arrow pointing up indicates a time delay after a relay has been energized, pointing down after it has been de-energized. Limit switches use a wedge directly on the contact arm symbol to represent the wedge-type levers that usually activate them. The foot switch has a pedal-like shape directly on the contact arm. Pushbuttons generally are spring loaded to their de-energized position and are drawn this way. Toggle and selector switches are drawn with a straight line to indicate the movable arm. Mechanical connections to switches are shown as solid lines if the connections are part of a single unit and dotted lines if they are located some distance away. See Figure 2-3.

Relay contacts resemble the electrical symbol of a capacitor. It should be remembered that the two parallel lines on a control diagram always represent a set of relay contacts and that a different symbol is used for the capacitor. Normally-closed contacts have a diagonal line through them to represent a closed position.

Indicating Lamps

Indicating lamps, (Figure 2-4) have a round symbol with rays to represent light coming from them. The color or function indicated by the lamp's being *on* is sometimes written inside the symbol.

Coils and Solenoids

Relay and motor starter coils are represented by a single circle with some notation to identify the function inside. Solenoids have a unique symbol to differentiate

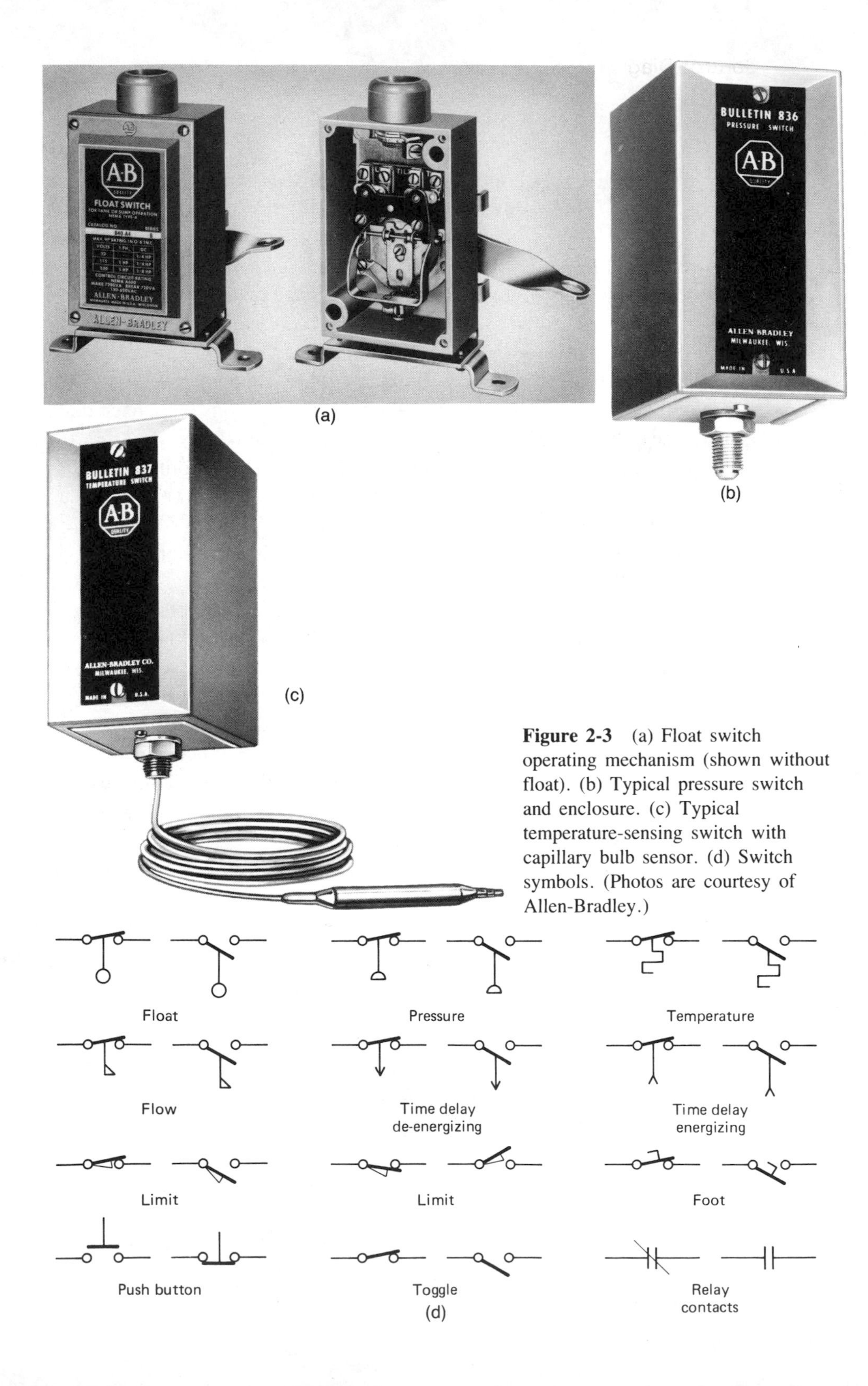

Figure 2-3 (a) Float switch operating mechanism (shown without float). (b) Typical pressure switch and enclosure. (c) Typical temperature-sensing switch with capillary bulb sensor. (d) Switch symbols. (Photos are courtesy of Allen-Bradley.)

Figure 2-4 (a) Indicating lamps. (b) Illuminated push button. (Photos courtesy of Allen-Bradley.)

them from coils. See Figure 2-5. Although the functions of the two devices are different, the symbols are sometimes interchanged. This is not a good practice, but you should be aware that it is done.

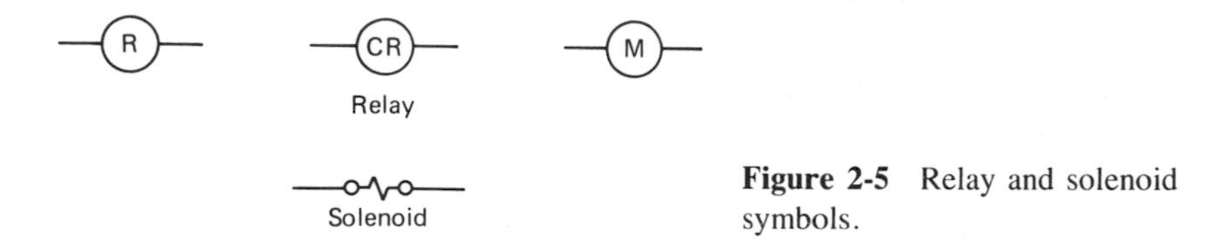

Figure 2-5 Relay and solenoid symbols.

CONTROL DIAGRAMS

The control diagram is supposed to be the simplest form possible to represent the functions of a control circuit. Basic parts of the circuit are a power source, the device under control or load, and interconnecting wires from the power source to the load. The type of power source depends upon the process under control. In large industrial plants, 480-V circuits are economical and are used for many types of equipment. Using 480 V for a control circuit is possible, and sometimes necessary, but it is not considered good control practice to do so. When high voltages are used, the insulating value of the equipment must be increased and this adds to the cost of that equipment. When high voltage sources "short out," a large amount of energy is expended, which may be very destructive at the location of the *fault*. High voltage faults therefore can be more destructive than lower voltage faults. Maintenance personnel must also work on these circuits. The danger from electrical shock while working on a low voltage source is less than that from a high voltage source. A common control voltage is 120 volts (AC). Since 120 VAC equipment is most commonly used, the price of parts is usually lower and the availability is also increased. This is an important consideration when a large number of parts are used or when replacement parts must be accessible within a short period of time.

The power source on a control diagram is represented in several ways. When a high voltage power source (480 V, e.g.) is reduced to the 120-V *control voltage*, a transformer is used. The symbol for a transformer is included on a control diagram to indicate a reduction from the power source voltage down to a control voltage. Within a large firm it may be a policy, or just common practice, to have all control voltages be a certain value.

To simplify the drawing of control circuits, two parallel lines are used to indicate the control voltage source. Typically these are labeled L1 and L2. Figure 2-6 illustrates a common method for representing the power source and the control voltage on a control diagram.

Each device, or load, in the control circuit will be connected to L1 and L2 in some manner. A load is indicated by using the appropriate control symbol on the diagram. These are usually placed on the right hand side of the diagram. Any controlling or *pilot* device is generally placed to the left of the control symbol. A guide to use in determining a load from a pilot device is to remember that a load must have the full potential of the control voltage to operate and that it consumes power. The pilot device is a switching or routing mechanism and generally does not consume power. Pilot devices exists as open and short circuits in their open and closed positions.

Interconnections between components are represented with lines to indicate the wires connected to each device. They are drawn to the symbol in a manner similar to the way the wiring is actually connected. Interconnecting wires, their symbols, and the control voltage symbols are all combined in such a way that they resemble rungs of a ladder. For this reason they are often referred to as *ladder diagrams*.

If a control voltage is understood to be a certain value, the symbol for the transformer is generally not included when smaller sections of the control diagram are the point of discussion. The transformer should be included any time there is a change in control voltage or a need to clarify information about the power source. When control circuits are designed, it is often more important to illustrate the *function* of the circuit than the control voltage (which may be decided at a later date or may be varied to fit individual specifications). Figure 2-7 illustrates a typical control diagram with all of the separate elements combined, but no control voltage specified.

Basic Diagram Rules

A few simple rules need to be understood for a circuit diagram to be functional when it operates. One of the rules is that at least one load device must be included

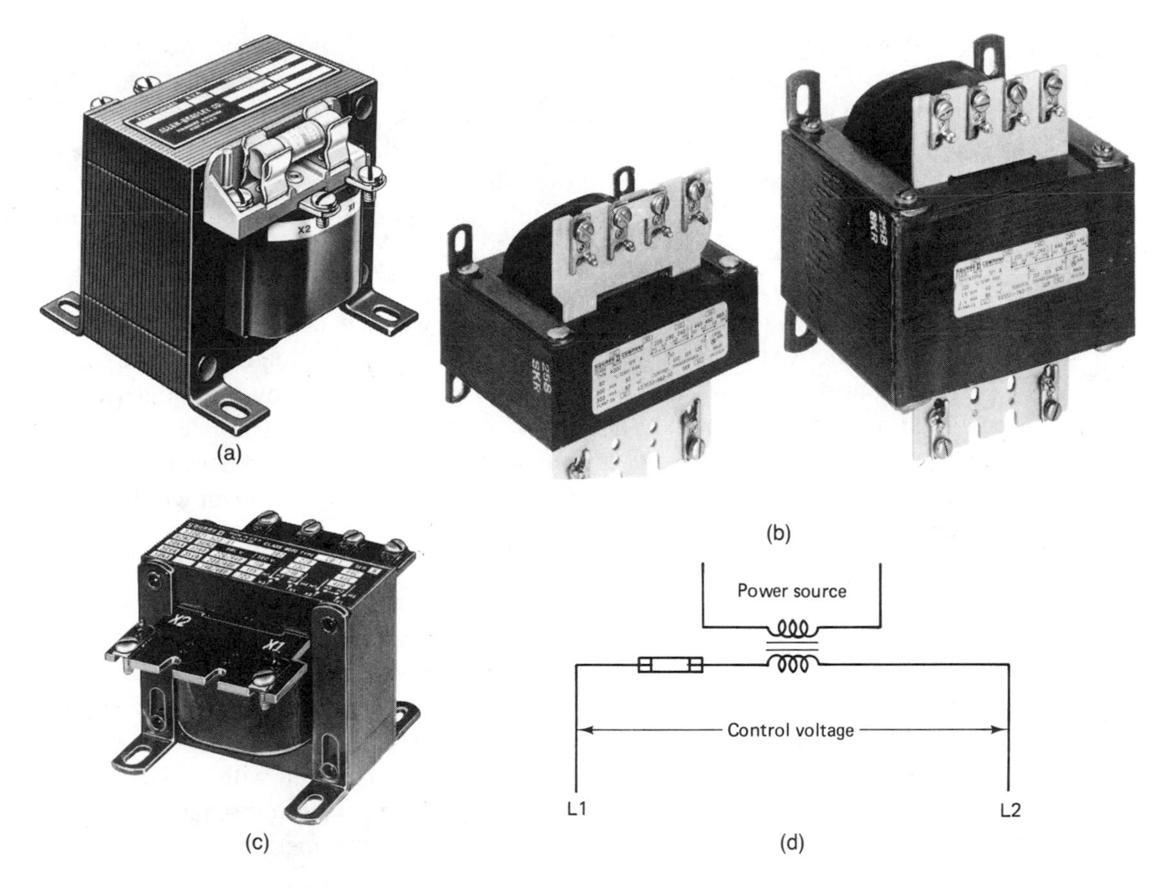

Figure 2-6 (a) Control transformer shown with built-in fuse (courtesy of Allen-Bradley). (b) Control transformers (courtesy of Square D Company). (c) Control transformer shown without built-in fuse (courtesy of Square D Company). (d) Control transformer schematic diagram.

in each *rung* (individual circuit) of the diagram. Without a load device the pilot devices would be switching an open circuit to a short circuit between L1 and L2. Short circuits are allowed within the pilot device section of the circuit as part of circuit design but must never exist directly across the control voltage source. The typical pilot device is a switch which has only two conditions, either an open- or short-circuit.

Figure 2-7 Typical control diagram.

A second rule is that additional load devices should always be placed effectively in parallel with the first load device. Effectively in parallel means that it is not possible to place the separate loads in series with each other by operating the pilot devices. Two 120-V load devices in series will not operate properly when placed across a 120-V control voltage. Each will have approximately 60 volts developed across it, and this would not be enough voltage to allow either device to function.

A third rule is that all additional *stop or off* functions must be placed in series. Figure 2.8 illustrates why this is true. Two switches are placed in parallel with each other to turn pilot light PL1 on or off. Either switch can be used to turn PL1 on, but if both SW1 and SW2 are closed, operating one switch individually would not allow the light to be turned off. A parallel path exists for current to flow and the light stays on. The correct way to connect the two switches to allow for a stop or off function is shown in Figure 2-9. Operating either switch will break the circuit and allow the light to go off.

A fourth rule is that all additional *start or on* functions should be placed in parallel. Violation of this condition is illustrated in Figure 2-10. Assuming either SW1 or SW2 is to be used to turn on PL1, operating only one switch will not allow this to happen. Both switches would need to be closed to turn the light on. The correct method, shown in Figure 2-11, is to place the switches in parallel. Operating either SW1 or SW2 will allow the light to come on.

Rules three and four are to allow additional remote functions to be added to circuits and to provide for safety considerations. Special combinations of both rules can be developed to further enhance the safety considerations of a particular circuit.

Two-Wire Control

Circuits illustrated so far have been what are commonly called *two-wire* control circuits. Two-wire control circuits generally describe a circuit that requires a wire from L1 to the pilot device and one more wire from the pilot device to the load, thus two wires. This really only applies in the most basic of circuits, since adding additional pilot devices will obviously increase the number of wires required. Figure 2-7 is a two-wire control circuit (a more general meaning is for manual control). In its simplest form a manual control uses a *maintained contact* device that must be manually moved from one position to another. The light switch in your home would be a typical example.

Three-Wire Control

Three-wire control describes the wiring requirements of a normal start-stop station in automatic control. Figure 2-12 illustrates using three-wire control and

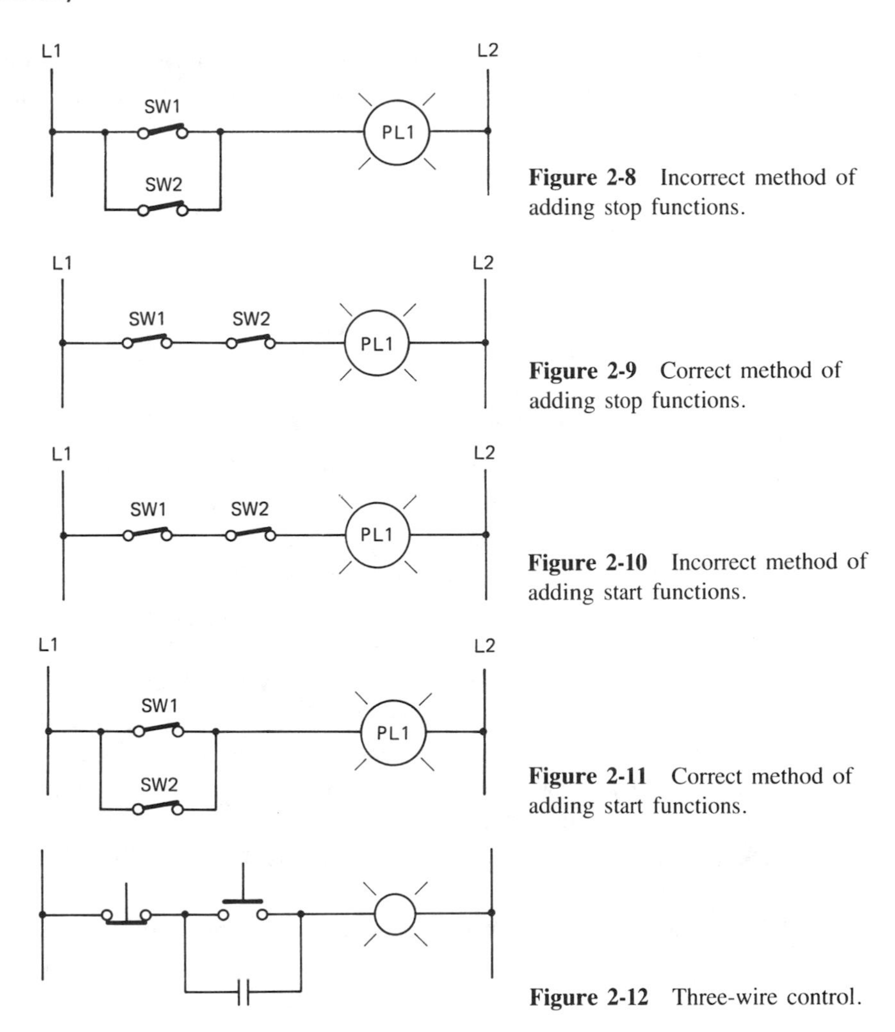

Figure 2-8 Incorrect method of adding stop functions.

Figure 2-9 Correct method of adding stop functions.

Figure 2-10 Incorrect method of adding start functions.

Figure 2-11 Correct method of adding start functions.

Figure 2-12 Three-wire control.

momentary contacts. One wire connects L1 to a stop button (in series) and another wire connects the stop button to a start button and a relay contact (both in parallel). This wire is "electrically" at the same point even though it appears to be two separate wires in the diagram. A third wire connects the opposite side of the start button and the relay contact to the relay. It is from this three-wire (electrically) arrangement that the circuit gets its name. The majority of circuits found in an industrial environment are the automatic or three-wire type.

SUMMARY

A thorough knowledge of electrical control diagrams is the key to understanding circuit operations. If you feel comfortable with the use of the symbols presented in this chapter, you will have little trouble designing simple circuits. Chapter 3 will enhance your knowledge and take you through step-by-step analysis of some common industrial circuits.

EXERCISES

On a separate sheet of paper, complete a response for each question, statement, or problem listed below.

2.1. What are electrical diagrams used for?

2.2. Why is a block diagram different from an electrical diagram?

2.3. What is the major function of the block diagram?

2.4. What are wiring diagrams?

2.5. Name the two types of wiring diagrams. Which is most commonly used?

2.6. Describe one of the major uses of a wiring diagram.

2.7. Two 25-lb rabbits accidentally jumped into two uninsulated power lines. One of the lines carries 480 VAC, the other 120 VAC. Which of the two lines would a rabbit be more likely to survive on?

2.8. Name the control element commonly used to step down a high voltage power source (e.g. 480 VAC) to a lower voltage power source (e.g. 120 VAC)?

2.9. Pilot devices do not consume power. Why?

2.10. For any functional circuit diagram, name two of the four basic rules that should be observed.

2.11. What would happen if two identical loads were connected in series with a 120-VAC power source?

3

Control Thinking

OBJECTIVES

Upon successful completion of this chapter, you should be able to

(1) list and discuss the reasons for *control thinking*
(2) describe the operation and functions of the start-stop circuit, the stop-jog circuit, and the start-stop-forward-reverse circuit
(3) formulate the criteria for selection and application of each of the circuits mentioned by the preceding objective
(4) list the safety implications of *control thinking*

INTRODUCTION

Control circuits are designed with a purpose in mind. It is important to realize the final goal or purpose of every circuit prior to its design. One way of doing this is through a process we shall call *control thinking*. Three simple ideas are involved.

Although they are typically involved with the development of logic circuits, they apply equally well to relay circuits. The process divides each circuit into an input section, a decision section, and the controlled device or "final element." First, you need to identify which are required inputs to the circuit. There should be some information or requirement presented to indicate a need for the circuit to operate. This input might result from the mental process of the machine operator in manual operation or from some automatic pilot device such as a float switch in automatic operation. The second section involves a decision to be made. Again this can be the operator or a pilot device. Finally the decision section either turns the final element on or off. Each of these ideas will be further developed along with the use of a ladder diagram. Some circuits can be divided into each section very neatly while others have the first two sections overlapping each other to some small degree or they entirely coincide with each other. As the number of components within a circuit increases, the importance of control thinking also increases. The following examples illustrate construction of ladder diagrams from control thinking.

Example 3-1

Let's suppose you were asked to design a circuit to turn a light on. The light is to be manually operated from one point and with minimum cost using standard residential voltages and equipment. These are the only design specifications given.

First, begin by doing a little control thinking. The input section must have a voltage source present, but it does not represent any information. The decision section is to be manually controlled by some operator. The output of the decision section will turn the light, the final element, on or off. This is a case where there is no data or signal of informational value to the first section. The decision and final element sections constitute the circuit.

Examine Figure 3-1 as each element is discussed. The power source will be 120 VAC, and it is represented by the two vertical lines L1 and L2. The "120 VAC" written between L1 and L2 is to show another way voltages are sometimes represented on ladder diagrams. The control symbol for a lamp (light) is drawn on the right hand side of the diagram. The general flow of a ladder diagram is from left to right with the final result of all actions taking place in the final element. For this reason the final control elements are usually placed on the extreme right hand side of the diagram. The pilot device chosen from the specifications given would

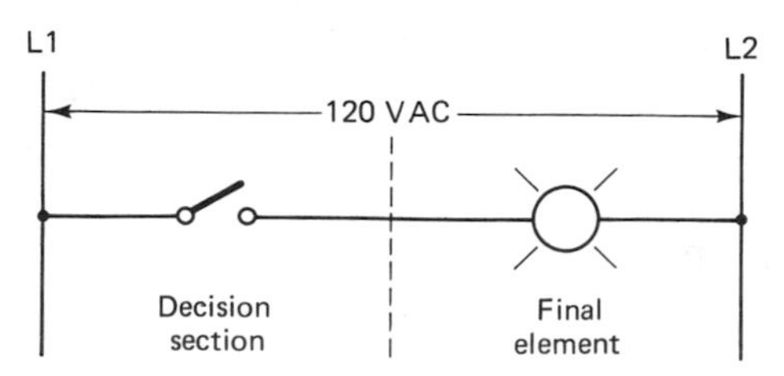

Figure 3-1 Light switch circuit.

probably be a simple toggle switch. (Is it a SPST or a SPDT?) The toggle switch is drawn on the left hand side of the diagram. Solid lines indicate interconnecting wires to the devices. Remember this is a functional diagram, not a wiring diagram. A vertical dashed line divides the circuit into the control thinking sections under discussion. Normally these lines and the labels are not placed on ladder diagrams. They are for illustration purposes only. Notice that neither of the symbols has a label to indicate which device it would be within a larger circuit.

This completed circuit is a two-wire control circuit. Does it satisfy the specifications given?

Example 3-2

Since you did so well in designing the first circuit, your skills are needed once more. The specifications that follow are given to you. Someone must open a valve on a tank of water once it reaches a certain level. Knowing it fills very slowly, there is no need to have someone stand and watch the water level as the tank fills up. A float switch is available from the supply area along with an indicator light. The desired operation is to have the float switch close and turn on the indicating light when the water level reaches a predetermined value. Since you know that the control voltage in your plant is standardized, it is not given in the specifications.

Examine Figure 3-2 as each section is discussed. The final control element will be the indicating light. When the water level reaches the float bulb, the float switch will close. This operation supplies information that the water level is high. There is really no decision to be made in this circuit. The information and final element complete the entire circuit.

Example 3-3

The water level circuit worked very well, but now a new pump has been installed to help drain the tank. The new specifications are given to you. The indicating light circuit is to be used to indicate a high water level. When the worker goes out to drain the tank, an electric pump is to be used. Since the operator must hold a button to operate the pump but cannot see or hear it, an indicating light should be installed to indicate when the pump is running. The operator is to hold the button until a second float switch turns the pump off. This is indicated by the light's turning off. In Figure 3-3 the final element part of the circuit has two devices in it—the pump starter, indicated by the circle, and the indicating light symbol with "pump" written inside. Notice that they are in parallel with each other. The float switch

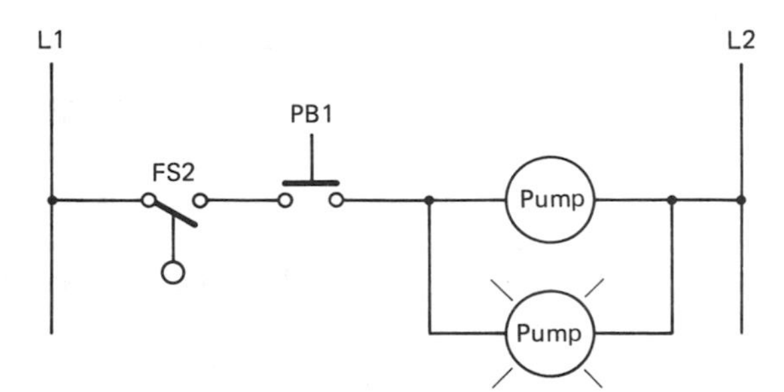

Figure 3-2 High level indicator circuit.

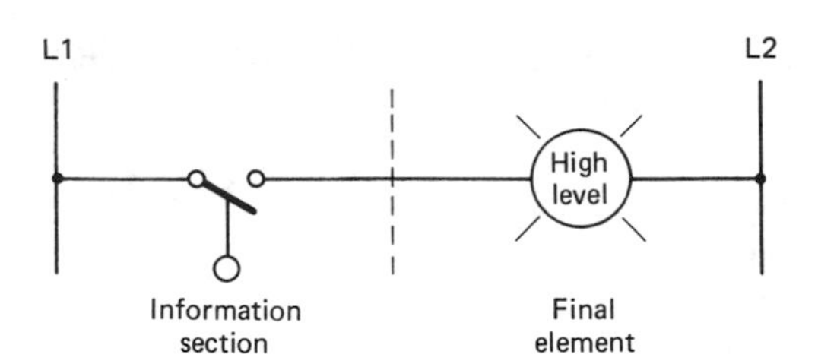

Figure 3-3 Pump control circuit.

again supplies the information that the water level is above a predetermined level. The push-button switch must be pushed by an operator. This requires a decision on the part of the operator and therefore is the decision section.

Figure 3-4 is the result of combining the two circuits. Notice that each symbol is now labeled. When a large number of symbols are used on a diagram, the diagram must indicate which device is represented by each symbol or confusion will result. The abbreviations FS1 and FS2 represent two separate float switches. The high level indicating circuit is on one rung of the ladder diagram and the pump circuit on a separate rung. The continuation of L1 and L2 indicates they use the same power source (control power). Each separate circuit represents an entire function in the diagram. An inoperative component in one function does not affect the operation of the other function in this example. More complex circuits do have interfunctioning components. That is, the malfunction of a component in one functional part of a circuit can affect another separate function.

The circuits given in Figures 3-1 to 3-4 are not the only possible nor the best circuits available. They do meet the minimum specifications given for their design. The specifications are very important for this reason. If no specifications

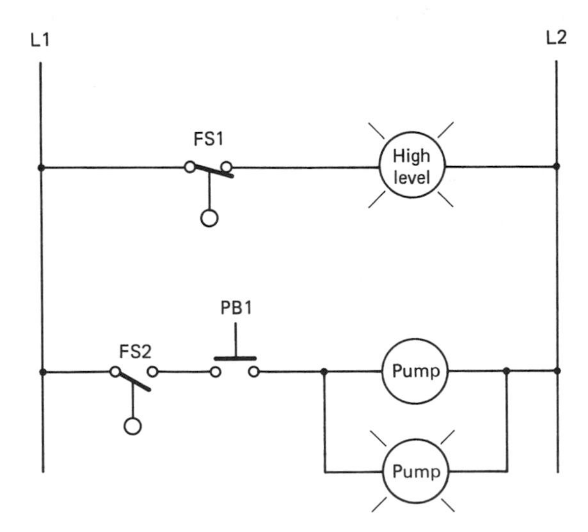

Figure 3-4 High level and pump control circuit.

are given, you should make an effort to find out what "exact" operation is expected from the final design. By knowing the function, you can logically think through the circuit to determine the components necessary. Many times a function can be thought of as a unit of components. When similar requirements are encountered, this part of the circuit will already be designed and can be put in place as a single unit. This will save time and aid in troubleshooting the circuit for malfunctions. You should also notice that the final element function must always be present, but the information or decision sections can exist individually, together, or sometimes with overlapping functions. Abbreviations used in diagrams are generally shortened forms of the symbols function. For example, FS is a float switch, PB is a push button, and CR is a control relay. These are typical but by no means the only ones used in the industry. Each engineer, designer, or firm has a preference for which symbols and notations they use. Many times they follow standard symbols, but once in a while they tend to personalize their diagrams and deviate from the accepted symbol. When in doubt about the meaning of a symbol or abbreviation, always ask someone to clarify it for you, if possible.

NORMALLY-OPEN AND NORMALLY-CLOSED CONTACTS

Electrical contacts that are operated by a relay are shown in their de-energized position on the relay. A *normally-open* contact will be closed when the relay operates and a *normally-closed* contact will open. It is very important that this is understood since this operation must be mentally visualized as you study the operation of control diagrams. Remember, all control diagrams are drawn in their de-energized position.

START-STOP CIRCUIT

A very common control circuit in industry is the start-stop circuit. Figure 3-5 illustrates the typical form and components used. Two push buttons, one normally open and one normally closed, are used with a relay, M. The two parallel lines below the start button represent a single set of contacts that are part of the relay. They are also labeled M to indicate this.

Figure 3-5 is shown in its power off, or de-energized position. With power available at L1 and L2, the relay will not function, since a complete path for current flow does not exist. This is because an open circuit is presented by the start button and the normally-open contact M. The stop button is normally closed, and the electrical potential from L1 is present through it to the left side of the start button and contact M.

If the start button is pushed, as shown in Figure 3-6, to the closed position, a complete path exists for current to flow to the relay. The relay will operate and contact M, which is part of the relay, will close. Look at Figure 3-7 and notice the start button shown as normally open. This is the condition it will return to after the operator releases the button. A dashed line is used to indicate the closure of contact M, its energized condition. (This is done for illustration purposes only.)

The circuit in Figure 3-7 is complete from L1 through the stop button, the closed contact of M and to the relay coil. Contact M is called a *holding circuit*, since it holds the relay in a closed position by providing a current flow path to it. Now if the stop button is pushed, as in Figure 3-8, the current path is broken and relay M will no longer operate. Since relay M is not operating, contact M will

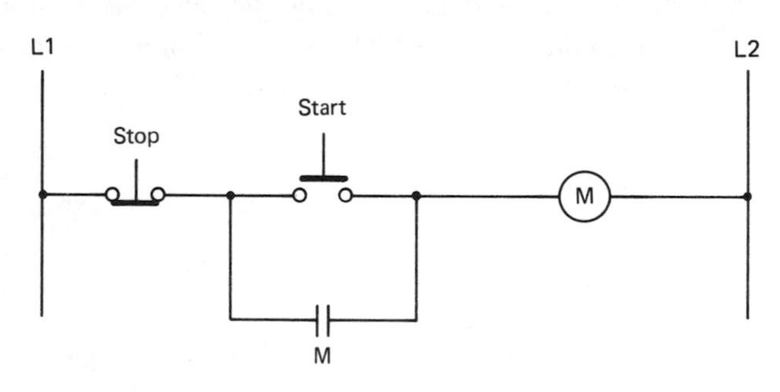

Figure 3-5 Start-stop circuit deenergized.

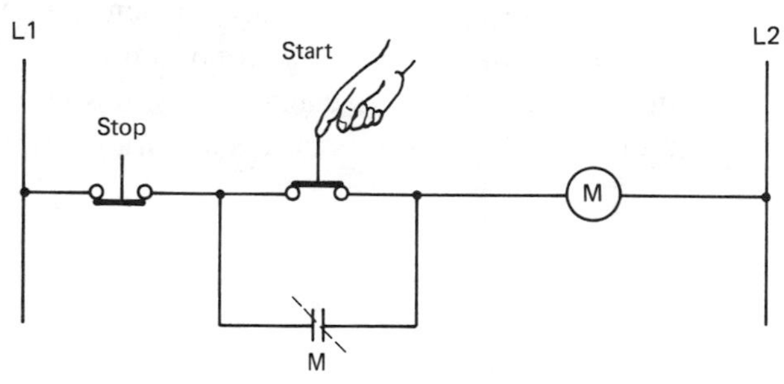

Figure 3-6 Start-stop circuit energized.

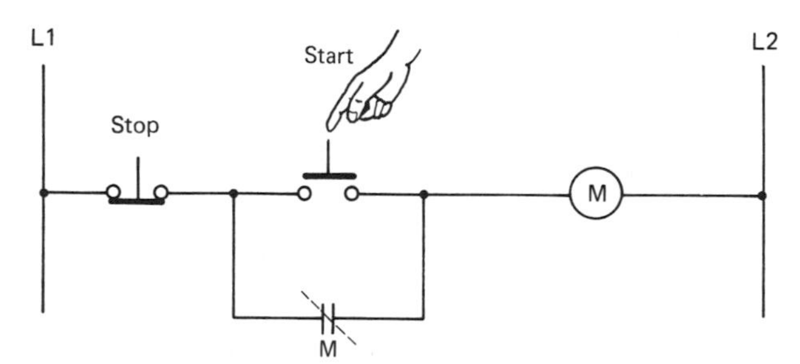

Figure 3-7 Start-stop circuit energized.

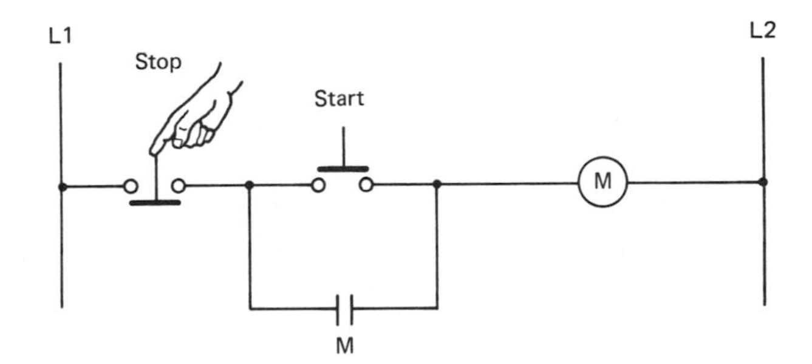

Figure 3-8 Start-stop circuit deenergized.

return to its de-energized position, which is normally open. The circuit now returns to the condition in Figure 3-5 described earlier.

A motor starter is a relay that switches the power on and off to a motor. Three separate sets of contacts are used in three phase loads, and one or two are used for single phase loads. In addition to the power contacts, one or more auxiliary or control contacts are built into the starter for use in control purposes. Another component of the starter is the *overload assembly*. The overload assembly detects excess heat developed by currents above the rated value of the "heaters." The overload unit controls an additional set of contacts that are normally closed. When an overload is sensed, these contacts open and break the current flow path of the control circuit. These contacts are usually labeled "OL" and are shown as a series of normally-closed contacts on the right hand side of the control diagram. Electrically they are in series with the stop button and thus serve as additional stop functions. Recall rule three given in Chapter 2.

Figure 3-9 shows a typical control diagram with additional stop and start buttons and overload contacts.

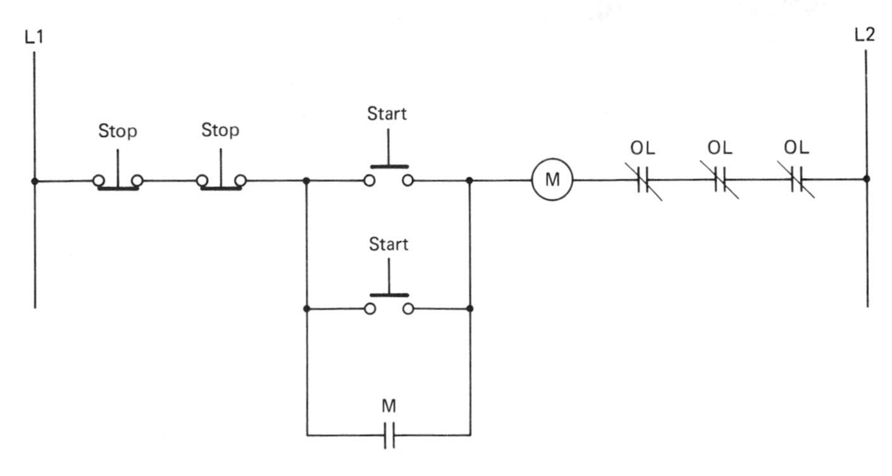

Figure 3-9 Typical start-stop circuit.

START-STOP-JOG CIRCUIT

Once the start button in Figure 3-5 is pushed, the relay will remain energized until the stop button is pushed. Sometimes it is desired to operate the relay momentarily without its remaining energized. This is called *jogging*. Suppose you have a motor that is run by a start-stop circuit. Occasionally you must run the motor and stop it without it running continuously. Figure 3-10 illustrates a circuit that would do the job. The start-stop function is basically unchanged.

Notice that the jog button is in parallel with the start button, but one portion of the button is in series with the *holding contact* (the contact that provides the holding circuit). When the jog button is pushed, a complete path is made for current flow to the relay. Once the relay is energized, contact M is closed; but as long as the jog button is held down, a complete path is not present through the holding circuit. If the jog button is released, the bottom contact of the button no longer provides a complete path and the relay de-energizes. For a brief moment neither contact of the jog button provides a complete path for either circuit. It is for this reason the relay remains de-energized. When the jog button is not being used, its upper contact provides a complete path for the holding circuit (if the start button is pushed and the start-stop circuit function operates normally). The solid line between the two contacts of the jog button indicates a mechanical connection between them on the same button. Because the manual operation of the jog button disconnects part of the circuit (intentionally and by design), this is a manual *interlock*. Interlocks are *safety, limit* or *sequence* control operations in a control circuit.

START-STOP-FORWARD-REVERSE CIRCUIT

Sometimes it is desirable to operate a motor in both the forward and reverse directions, (but not at the same time). An overhead crane is a good example. In this case up and down can be substituted for forward and reverse. Figure 3-11

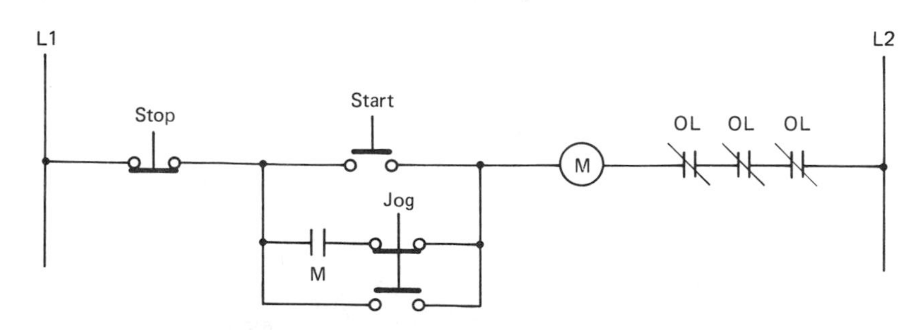

Figure 3-10 Jog circuit.

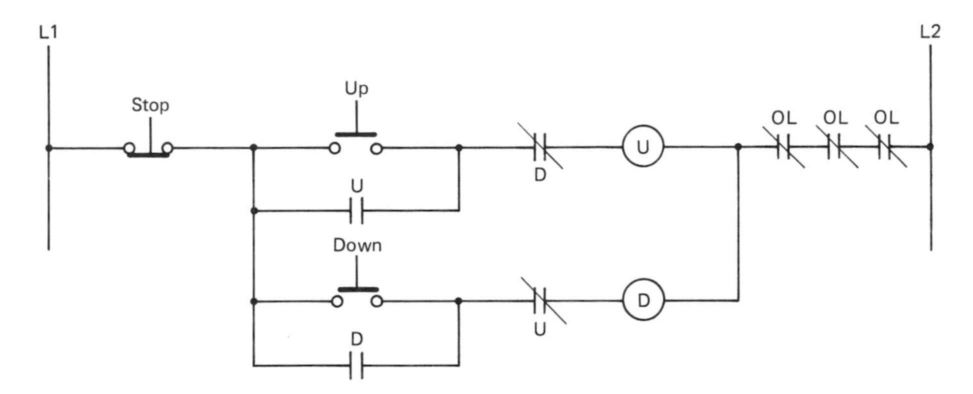

Figure 3-11 Up-down (forward-reverse) circuit.

shows a typical circuit. A separate relay is used for the *up* and *down* functions; therefore two relays are shown in the diagram.

The *up* function is a complete function within itself, and it occupies one rung on the control diagram. For the same reason, the *down* function is on a separate rung. The diagram is a start-stop circuit in each case with one exception. There are electrical contacts in each holding circuit that operate with the function opposite that being used. Let's examine the normal operation of the circuit.

It is desired to raise a load off the floor, so we push the up button on our overhead crane. Current flows through the normally-closed stop button, the up button being held down, and to the up relay U. When the relay U is energized, contact U closes, and current also flows through the normally-closed D contact (if relay D is de-energized). The holding circuit for the up function is now complete, and the motor will continue to run when the up button is released.

Now what would happen if we inadvertently pushed the down button while the crane was raising the load? Looking at the circuit in Figure 3-11 allows us to determine that the following conditions exist. The stop button is closed, both the up and down push buttons are open, the up relay is energized and the down relay is de-energized. The normally-closed D contact is still closed, the normally-open D contact is open. The normally-open U contact is now closed, and the normally-closed U contact is now open since relay U is energized. Even though we have pushed the down push button, the open circuit created by the normally-closed contact of relay U prevents current flow to relay D. These contacts that prevent inadvertent operations that could damage equipment are called *electrical interlocks*. To stop the crane, we can push the stop button. All current must go through the stop button, so if it opens, the entire circuit is de-energized. The down function operates in a manner similar to the up function. A single stop button is used in this circuit because the same motor is used for both functions and for the same reason only one set of overloads are used per motor.

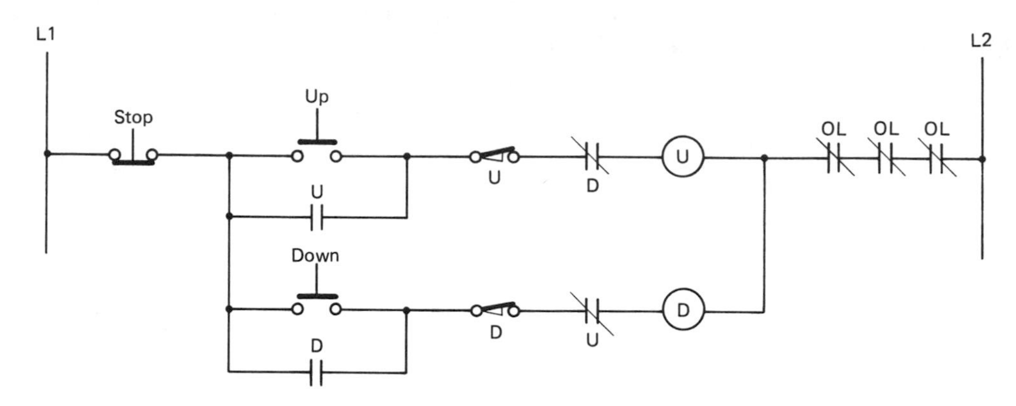

Figure 3-12 Typical crane-hoist circuit.

Figure 3-12 illustrates a more typical crane hoist circuit. To keep the cable from binding or rolling off the spool when the hoist reaches the upper and lower extremes of its operation, *limit switches* are installed. When the hoist reaches the upper limit of its travel, the UP limit switch is opened, and it functions as a stop button in the up circuit. When the lower limit is reached, the down function is de-energized by the DOWN limit switch.

SUMMARY

Learning to think in a logical sequence with a control circuit is very helpful when repairing electrical equipment. You should learn to visualize each circuit you work on and consider the function it was designed to control. Isolate each section until you locate a function that is not working. Then, using the function, isolate the faulty component in the smaller circuit. Chapter 4 will use typical circuits and then modify them to perform additional functions. A step-by-step analysis of the circuits will give you the opportunity to examine events as they take place and the resulting effects on other parts of the circuit.

EXERCISES

On a separate sheet of paper, complete a response for each question, statement, or problem listed below.

3.1. Name the three major sections of any control system or device.

3.2. How do these three sections each relate to control thinking?

3.3. Why might one defective component in a section of a control device affect another section of the same device?

3.4. What minimum information do you need to design a system?

3.5. In what position would a normally-closed contact and a normally-open contact be when they are energized?

3.6. Why is the jog button in a start-stop jog circuit sometimes called a manual interlock?

3.7. Give an example of a start-stop-forward-reverse circuit.

3.8. What would happen if the stop button is *open-circuited?*

3.9. What is the major function of an electrical interlock?

3.10. What function and/or limitation is common to an automobile transmission as well as to a start-stop-forward-reverse circuit?

4

Modifying Circuits

OBJECTIVES

Upon successful completion of this chapter, you should be able to

(1) distinguish between the redesign and modification of circuits
(2) list the reasons for modifying circuits, instead of redesigning them
(3) read and analyze circuit specifications
(4) discuss the operation and function of an interlock circuit
(5) modify a simple control circuit to perform a basic task

INTRODUCTION

The simple start-stop circuit is sufficient for many applications, especially if the function operated can be turned off or on independently without affecting any other process. Many times, however, it may be desirable to place additional requirements on the operation of a circuit. Once a specific requirement has been identified, it can be incorporated into the basic circuit. This chapter will illustrate how a basic circuit may be modified as new restrictions are placed on a process. Try to visualize the basic circuit or function that takes place in each stage of development and how individual functions are combined to form a single larger function. The example used will be a series of conveyors. Read the specifications for each example and then analyze the circuit that is developed from them.

CIRCUIT ONE SPECIFICATIONS

Three separate conveyors are to operate independently from each other. Each is to have three-wire start-stop controls and motor overload protection. A common control power source may be used, if desired.

The circuits in Figure 4-1 meet the specifications given. Each circuit may operate its conveyor motor with no interference to other conveyors. In this example each circuit is a basic start-stop function. The notations, C1, C2, and C3, represent the motor starter coil for conveyors one, two and three respectively. Contacts that operate when a coil is energized are labeled with a corresponding notation. Each component serves a definite function. Since all three conveyors have the same basic circuit, only one will be discussed.

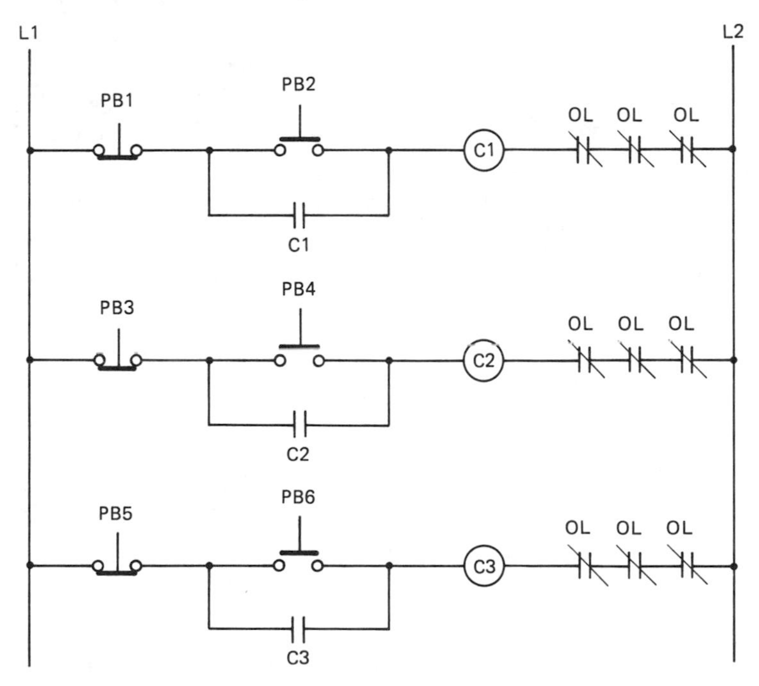

Figure 4-1 Start-stop circuits for conveyors.

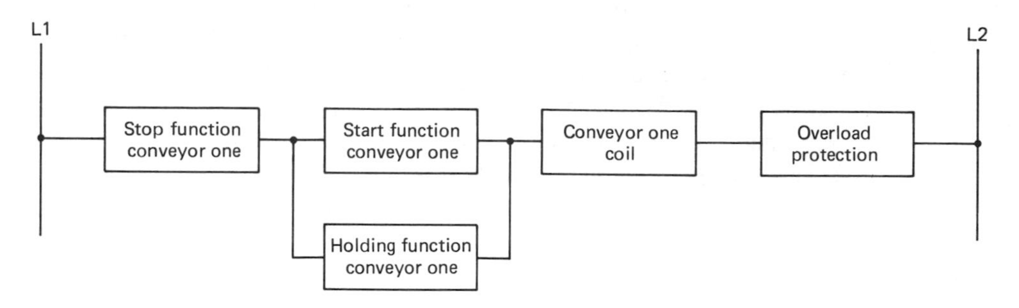

Figure 4-2 Functional diagram for conveyor one.

In Figure 4-2 the purpose of each component is listed by the function it serves. Examine each individual function and then consider the entire circuit as one whole function. Learning to think in functions can be very beneficial when troubleshooting or locating faults in circuits. Diagrams of this type are not usually developed, but this chapter will use them to help you visualize what each component does in a circuit. Motor overload protection is a standard practice and in future circuits will be included on all motor starters without explanation.

CIRCUIT TWO SPECIFICATIONS

The existing three conveyors are to be interconnected to allow for a sequential start. Conveyor three is to start first, followed by conveyor two and then conveyor one. If a conveyor does not start, the remaining conveyors must stay off. If a conveyor stops, all other conveyors that sequence after it must also stop. Except for the above conditions, each conveyor is to have individual start-stop control.

Circuit 4-3 is designed to meet the conditions given. As you examine the circuit, look for the component used for satisfying each specification. Mentally operate each circuit to see if it would fail to meet any of the specifications. If at any time one of the specifications is not met, then the circuit must be redesigned.

Let's go through the circuit step by step to see how it operates. (See Figure 4-3.)

1. Starting with PB2, if we try to start conveyor one (conveyor C2 and C3 are off), nothing should happen. Contact C2 is open and a complete path is not available to energize coil C1.
2. If we push button PB4, conveyor C2 will not run because of the open contact C3.
3. If we now push button PB6, conveyor C3 will start. Conveyor C3 is now running.

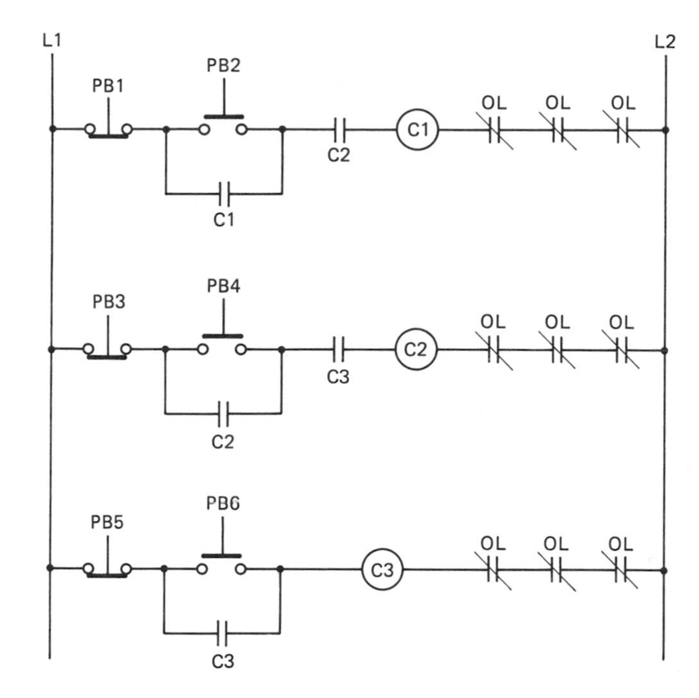

Figure 4-3 Sequential conveyor circuit two.

4. Starting again with PB2, conveyor C1 will still not run. Contact C2 is still open.
5. When PB4 is pushed, conveyor two starts because the running of conveyor three has closed contact C3, allowing coil C2 to be energized.
6. Returning one more time to PB2, we find that conveyor one will now start because contact C2 is closed when coil C2 was energized.

Review the specifications to see which conditions have been met. The only condition not tested so far is for a conveyor to stop and the remaining conveyors that sequence after it to also stop. Assume all of the conveyors are running.

7. Pushing PB3 will stop conveyor two by de-energizing coil C2.
8. Contact C2 will open and this will de-energize coil C1. If conveyor C2 stops, then conveyor C1 also will stop.
9. If PB5 has been pushed, then all three conveyors would stop.

Examine the circuit again and visualize what would happen if PB1 were to be pushed. Have all the conditions specified been met?

The specified conditions have been met. The functional diagram in Figure 4-4 for this circuit again illustrates how one large system can be subdivided into individual functions.

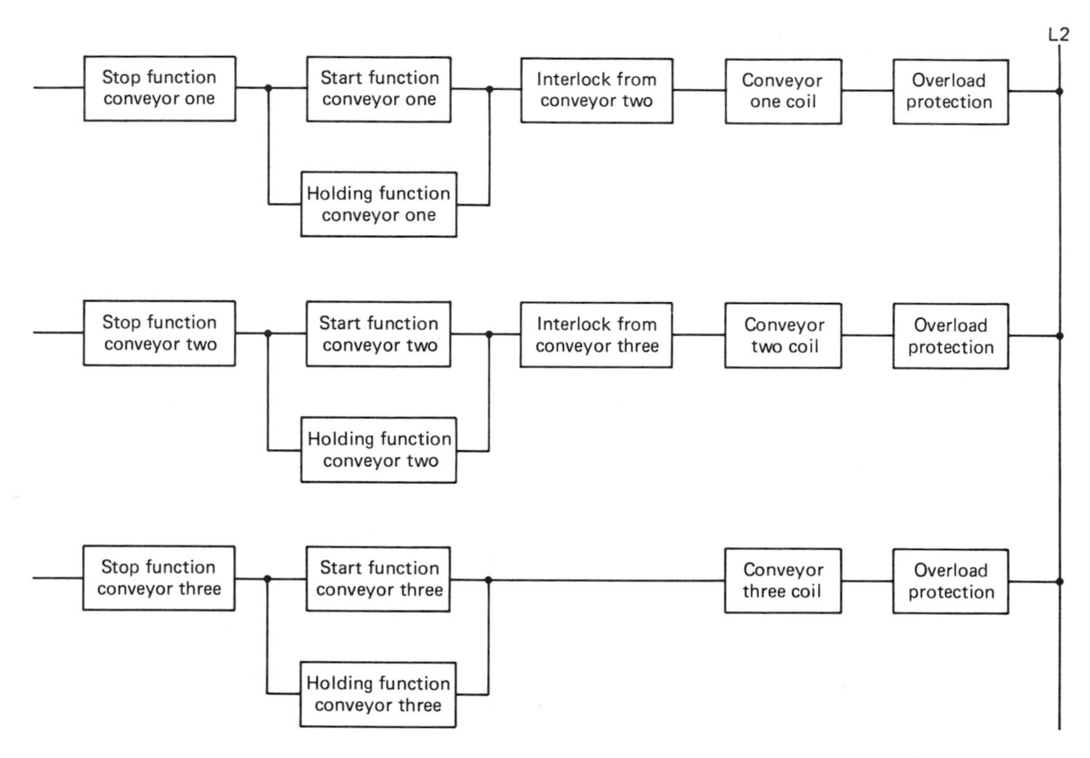

Figure 4-4 Functional diagram circuit two.

CIRCUIT THREE SPECIFICATIONS

Three conveyors are to start from one start button and sequence start, beginning with conveyor three. A time delay of 30 seconds is to elapse between each conveyor's starting. A stop button is to be provided at each conveyor to be used both for normal and emergency stops. If a conveyor should stop, all sequential conveyors shall also stop.

This is a major change from the previous circuits. Individual control is no longer specified and a time delay is to be added.

The circuit in Figure 4-5 was developed to satisfy this new set of specifications. Let's begin with the changes in the rung that controls conveyor C3.

In order for conveyor C2 to operate after 30 seconds, some means of timing must be initiated when conveyor three starts. Time-delay relay one, TD1, serves this purpose. This relay is a *time-delay on energizing* (TDOE) unit. This means that after 30 seconds the contact unit of the relay will operate—all normally-open contacts will close and all normally-closed contacts will open.

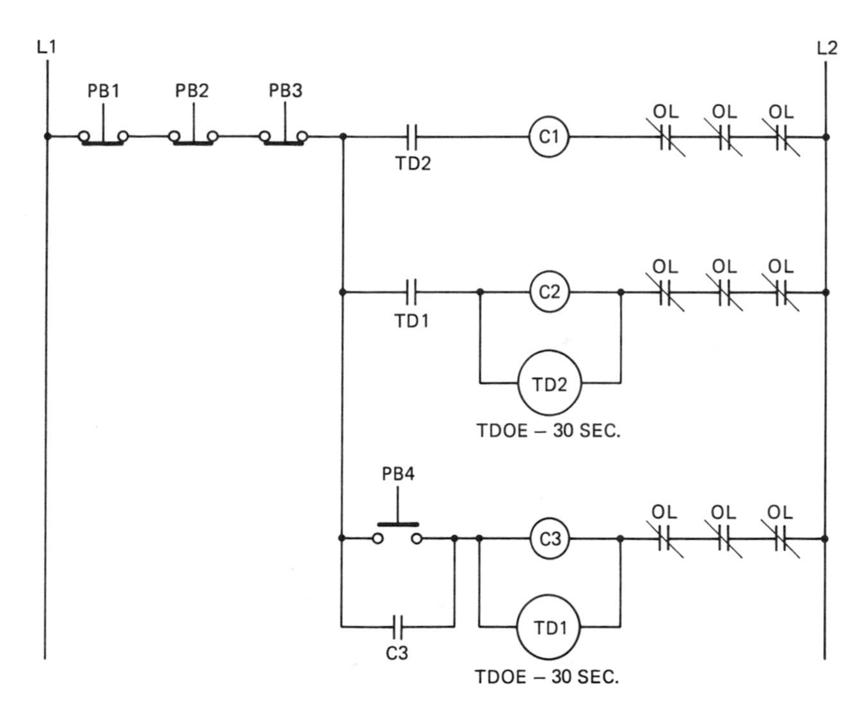

Figure 4-5 Sequential start with time delay.

1. The normal sequence of operation will be for PB4 to be closed to energize coil C3 and at the same time TD1.
2. After 30 seconds, contact TD1 in rung two will close allowing coil C2 and TD2 to energize.
3. After 30 seconds contact TD2 in rung two will close and coil C1 will energize.
4. All three stop buttons are placed in series and pushing any one will de-energize the entire circuit.

All the specifications have therefore been satisfied. Figure 4-6 illustrates this circuit in a functional diagram format.

Example 1

THE PROBLEM. Suppose we have eight (8) conveyors which start from one start button, in the order of 1 to 8. If a time delay of 30 seconds is to elapse between each conveyor's starting, how long would it take before the sixth conveyor starts? How long would it take for all of the conveyors to start?

Solutions Before the sixth conveyor starts, it would be necessary for each of the preceding conveyors to begin:

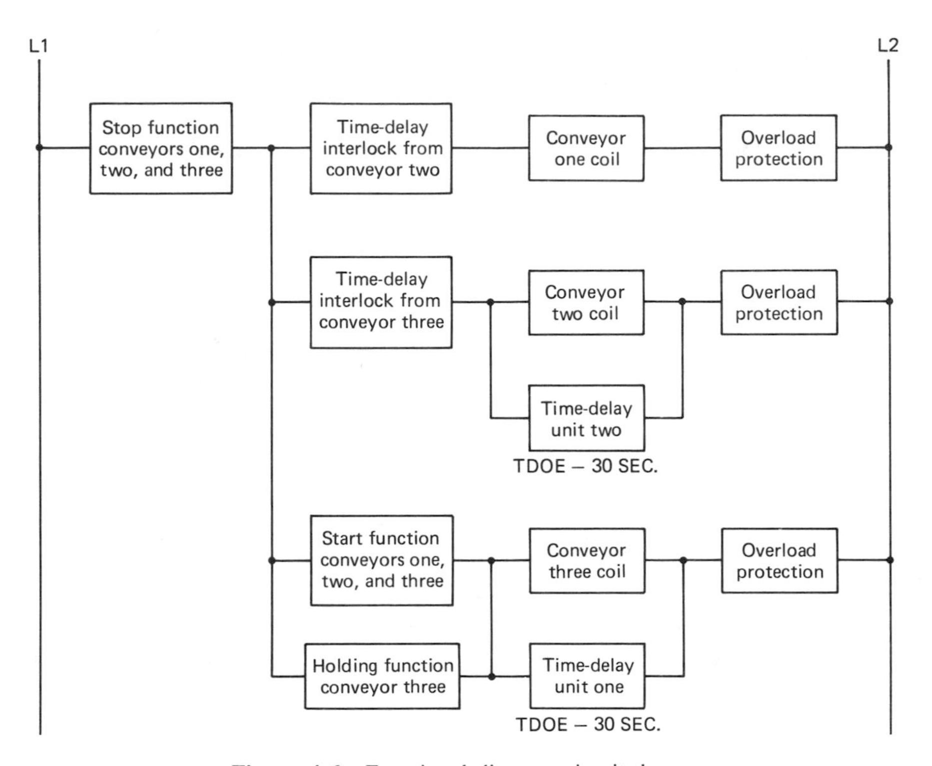

Figure 4-6 Functional diagram circuit three.

$$(T_1 \text{ to } T_2) + (T_2 \text{ to } T_3) + (T_3 \text{ to } T_4) + (T_4 \text{ to } T_5) + (T_5 \text{ to } T_6)$$
$$= 30 \text{ s} + 30 \text{ s} + 30 \text{ s} + 30 \text{ s} + 30 \text{ s} = 150 \textit{ seconds.}$$

It would take an additional 60 seconds for the 7th and 8th conveyor to begin:

$$(T_6 \text{ to } T_7) + (T_7 \text{ to } T_8)$$
$$= 30 \text{ s} + 30 \text{ s} = 60 \textit{ additional seconds.}$$

Therefore, the total time before *all* of the conveyors start is the time from the first conveyor *to* the eighth conveyor (T_1 to T_8), which would be 150 seconds plus 60 seconds, or 210 *seconds total*.

CIRCUIT FOUR SPECIFICATIONS

In addition to the conditions specified in circuit three, sets of photoelectric eyes are to be added to conveyors two and three. If photoelectric eye three is blocked by material on conveyor three, then conveyor two is to stop until the material in front of the photoelectric eye has cleared. A five-second delay is to be set in the photoelectric eye to give time for the normal flow of a material past the photoelectric eye without stopping the conveyors. Conveyor two is to have a similar unit to control conveyor one. Conveyors one and two are to start again automatically once the photoelectric eyes are cleared.

This may seem like another major change, but only two new components are added to the existing circuit to make the change. Photoelectric eyes have power applied to the unit at all times. The position of the light beam and a reflecting unit determines what makes it operate. The presence or absence of reflected light causes a relay contact to open or close within the unit.

Figure 4-7 shows the method of connecting the photoelectric eyes, PC1 and PC2, in our circuit. The timing function of the photoelectric eyes is similar to that of the time-delay relay. If the light beam is present with no material to block it, the time-delay relay is de-energized. When a box or other material blocks or

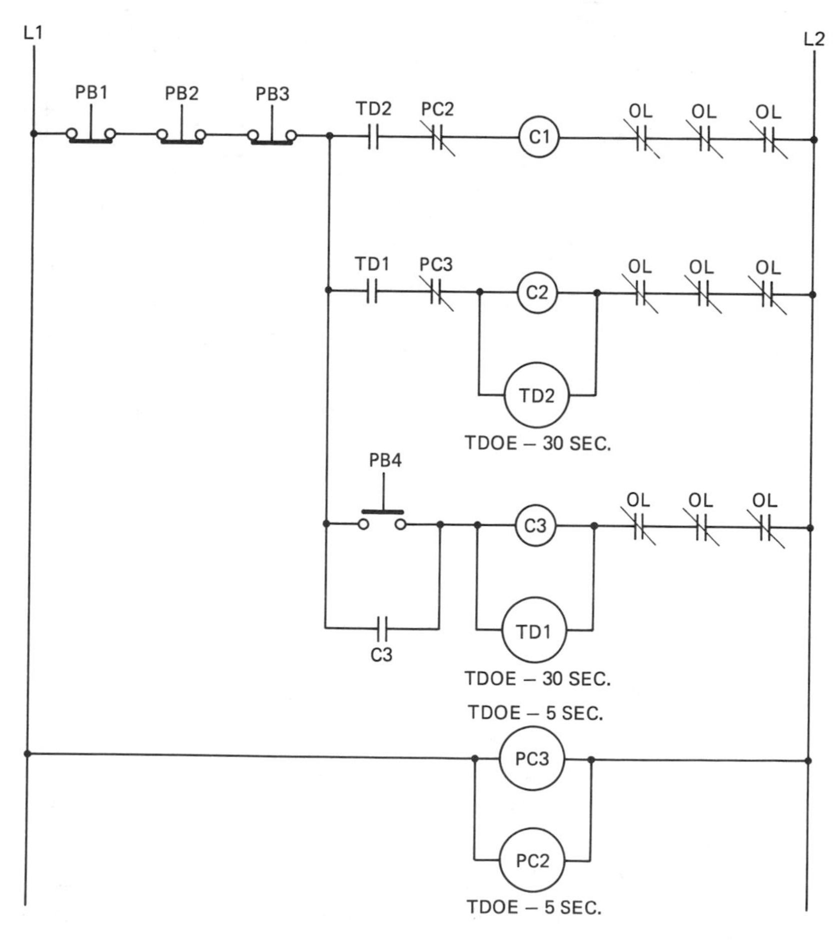

Figure 4-7 Sequential start with photoelectric eyes.

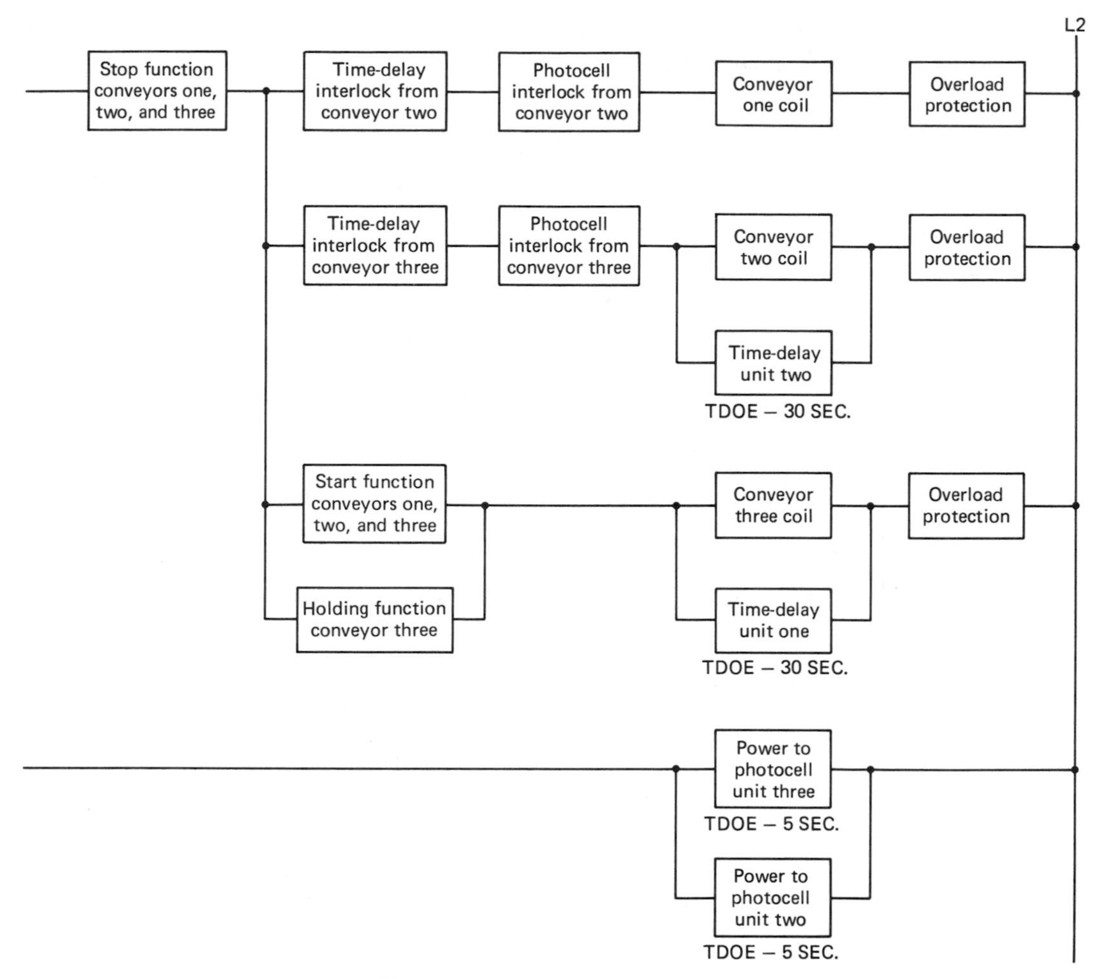

Figure 4-8 Functional diagram circuit four.

breaks the light beam, the relay energizes and five seconds later the contacts open or close. With this in mind let's examine the operation of the circuit.

1. All conveyors are off and PB4 is pushed.
2. Coil C3 and TD1 are energized.
3. In rung two, contact TD1 closes after 30 seconds, contact PC3 is closed if the light beam of photoelectric eye PC3 is not blocked, and coil C2 and TD2 energize.
4. In rung one, contact TD2 closes after 30 seconds, contact PC2 is closed if the light beam of photoelectric eye PC2 is not blocked, and coil C1 energizes.

5. A box in front of PC3 for longer than 5 seconds will cause contact PC3 to open and de-energize coil C2 and TD2. This will in turn cause TD2 to open and de-energize coil C1.
6. Once PC3 is clear, conveyor two will start again and after 30 seconds allow conveyor one to start.

Figure 4-8 illustrates the functional diagram of this circuit.

INTERLOCKS

Circuits two through four utilize what is commonly called *interlocks*. Interlocks allow or prevent some function from happening until specific conditions are met. Contacts C2 and C3 in circuit two, TD1 and TD2 in circuit three, and TD1, TD2, PC2, PC3, of circuit four are all interlocks. They may be normally open or normally closed, depending on the function they must perform.

SUMMARY

This chapter went into much detail to show how a system may be modified from a simple circuit to a more complex circuit. Each new modification followed the basic rules for adding devices to the circuit. If they were not obvious to you, please review the rules once again. Interlocks are one of the many ways of incorporating conditional requirements into a circuit. You should study each of the circuits until you fully understand it. A fundamental knowledge of these examples will enable you to take a *small functional unit* from a major circuit and apply it to a larger circuit.

Chapters 1 to 4 should supply you with a working knowledge that will allow you to develop functional circuits for most needs. The remaining chapters are designed to fine tune these basic skills. It will help to remember that there are numerous different possible circuits that can be developed to do a single task. Many times different people will design entirely different circuits for the same application. Each circuit should be evaluated for a given situation and set of specifications. In most cases, the circuit with the fewer components would be more practical.

EXERCISES

On a separate sheet of paper, complete a response for each question, statement, or problem listed in the following section.

4.1. List three reasons why a circuit might be modified.

4.2. List two reasons why the individual subsections of a circuit should be identified by function.

4.3. Four (4) conveyors start from one start button. A time delay of 45 seconds is to elapse between each conveyor's starting. If the sequence begins from conveyor one to conveyor four, how long would it take before the third conveyor starts? How long will it take for all of the conveyors to start?

4.4. What is the most common device used for providing time delay called?

4.5. One common term used with relays is TDOE. What does the term mean?

4.6. You were to design a delay system to provide time delay of 150 seconds. You have only two time-delay relays; one provides 100 seconds of delay and the other provides 50 seconds of delay. How would you go about designing for the 150-second delay?

4.7. What is an interlock?

5

Relay and Solid-State Logic

OBJECTIVES

Upon successful completion of this chapter, you should be able to

(1) describe differences between solid-state logic and relay logic
(2) discuss the operation of AND circuits, OR circuits, NAND circuits, NOR circuits, and NOT circuits
(3) describe the operation and application of memory gates
(4) demonstrate the purpose of a time-delay function
(5) illustrate uses of solid-state logic gates for industrial applications
(6) interface gates of one family with gates of another family and
(7) with reference to gate logic, describe what is meant by such common terms as *noise immunity, fan out, sinking, sourcing*, or *transition region*

INTRODUCTION

You may think devices, such as transistors, solid-state relays, and programmable controllers, are replacing all of the old equipment in use today; but this is not true. Many firms do not make rapid changes to new systems. Relay circuits are the most common control systems found in industry today. The flexibility they offer accounts for some of this popularity. If a system remains cost-effective, it may not be upgraded. As equipment does become outdated and difficult to replace, this is when upgrading becomes most economical. Some firms still prefer the simplicity of relays, and when they do purchase new equipment, they may mix some old equipment with the new technologies. The term used to describe this mixture of the old and new is called a *hybrid* system.

Conventional relays, limit switches and other pilot devices are often used as input controls to transistor circuits. One transistor device used is the *logic module*. Logic modules are low-voltage, low-power, decision-making circuits that may be connected to amplifiers to drive high-power control components. This chapter will discuss logic modules that plug into relay sockets. New advances will eventually make even these components obsolete, but the function they serve is incorporated into the newer advanced equipment; therefore, they will be used to introduce you to logic systems as control devices. To help the transition from relay control to solid-state control, logic circuits designed with relays will be presented first.

DECISION-MAKING CIRCUITS WITH RELAYS

AND Circuits

Specifications that require one function to be accomplished and a second function also to be completed either at the same time or as a precondition for another event to take place describe the AND circuit. This circuit can be formed by placing relays in series. Figure 5-1 illustrates an example of a safety interlock circuit found on many stamping machines. The control relay, CR, must be energized

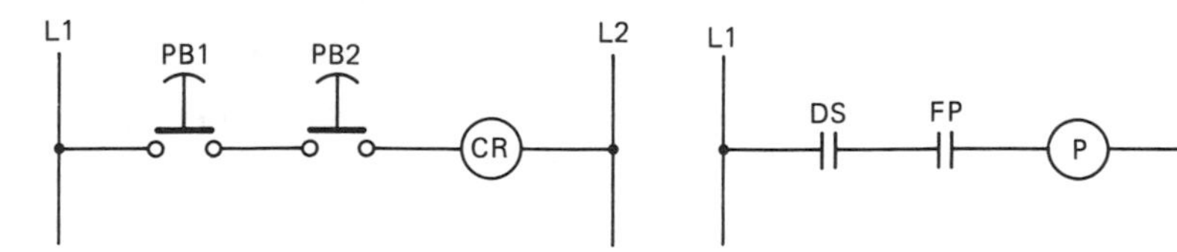

Figure 5-1 Safety interlock circuit (AND circuit).

Figure 5-2 Safety circuit for compactor (AND circuit).

before other parts of the circuit, which uses the interlock relay contacts, can perform functions (such as a hydraulic ram or a cutting die). Push buttons PB1 and PB2 are physically placed in such a manner that the operator cannot be near the ram or die and still push the buttons. At the same time, a prerequisite for ram movement would be that CR be energized. This example used push buttons, but relay contacts are also used to perform similar tasks. Examine Figure 5-2 for a hydraulic pump on a compactor, and notice the conditional requirement that both contacts DS and FP must be closed before pump P will operate. In this example the contacts will be closed when the operator is standing on a pressure-sensitive foot pad, FP, and the access door is in a closed position (indicated by switch DS). It should be noted that standard symbols are sometimes replaced with normally-open or normally-closed contacts, and abbreviations are placed above them to indicate their function.

OR Circuits

Another useful configuration is the OR circuit. The specifications require that either one condition or another separate condition can cause an event to take place. The example in Figure 5-3 shows a circuit than can turn a light on if a photocell, PC, senses darkness OR someone turns a light switch, LS, to the ON position. Note that either PC or LS can allow the light to be illuminated and that they can occur independent of each other. This is the basic criteria for the OR circuit.

NAND Circuits

Not all conditions will easily fit the simple circuits just described. It is sometimes necessary to specify that conditions NOT be present prior to circuit operation. This is done by using normally-closed contacts. Assume an object is on a platform, and it must be moved prior to the next operation of a machine. Two limit switches, LS1 and LS2, are closed whenever the object is on the platform. A light is turned off to indicate the object is not present. This may also be stated as "If LS1 AND LS2 are NOT cleared, the light will be on." Our example is

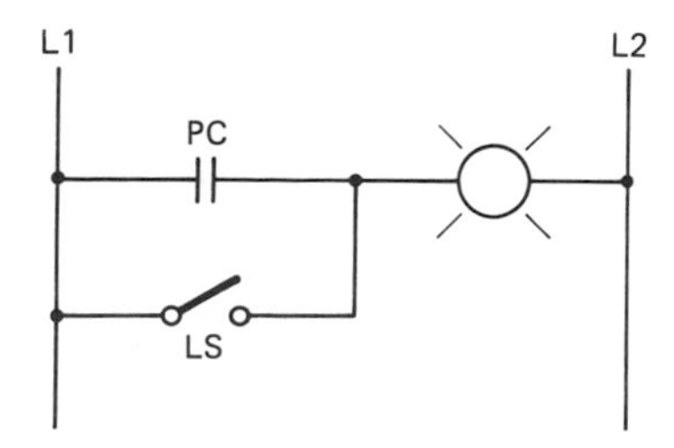

Figure 5-3 OR light circuit.

illustrated in Figure 5-4. The basic specifications to satisfy a NAND circuit are that if any or all input conditions are NOT present, an input will result.

NOR Circuits

A circuit in which the absence of any of the inputs will allow the event to take place is called a *NOR circuit*. Figure 5-5 is an example of a NOR circuit. If CR1 OR CR2 are NOT energized, the light will be illuminated. The specifications for a NOR circuit are that if all of the inputs are not present, there will be an output.

NOT or Inverting Circuits

There are occasions when an event is taking place that some indication is desired to specify the negative or opposite indication. This may be having a light come on and another go off, or alternate, with neither being on at the same time. Figure 5-6 illustrates a possible circuit that would indicate the negative of the input switch position.

Memory with Relays

There are two possible circuits that can be used to remember an input that has previously occurred. One method is to use a maintained-contact switch as in Figure 5-7. In this circuit, if the power is de-energized and then energized, the relay will operate again as if it remembered the last switch position. The circuit in Figure 5-8 illustrates another example; but in this case de-energizing the circuit

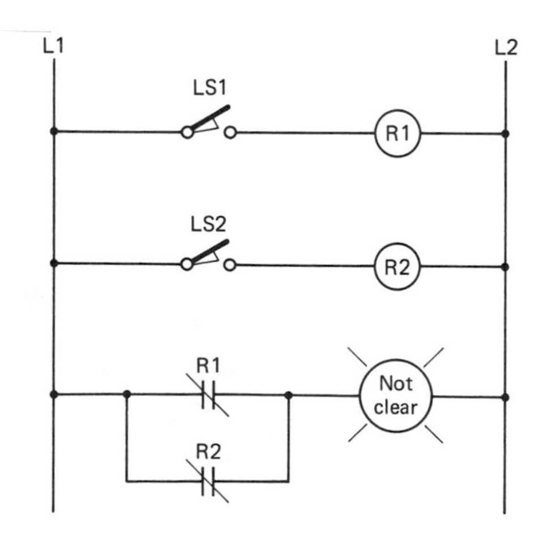

Figure 5-4 Position indicating circuit (NAND circuit).

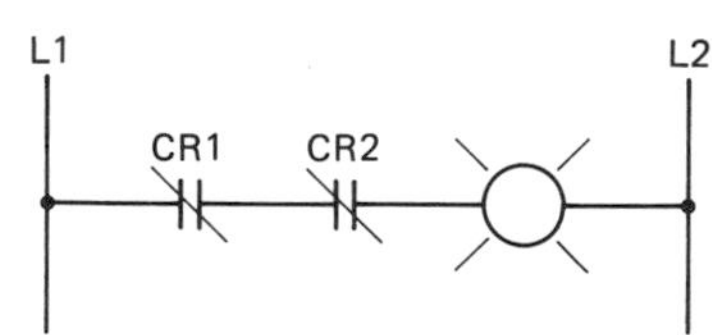

Figure 5-5 NOR circuit.

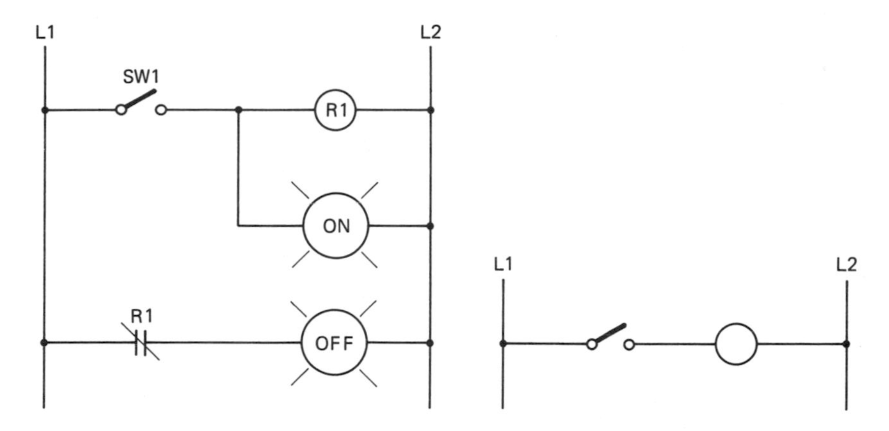

Figure 5-6 NOT or Inverting circuit.

Figure 5-7 Nonvolatile memory circuit.

will cause the relay to forget the last position the push buttons were in and not operate. Obviously, relays do not have a capacity to think or remember, but memory is often used in logic circuits to indicate the ability to recall or store past events.

Memory that is lost with a power failure is called volatile memory. Memory that is retained with a power loss is called nonvolatile.

Time-Delay Circuits

Time delays are incorporated using time-delay relays. These relays have been discussed previously, but it is helpful to recall that there are time delays available on energizing and de-energizing, and some units have both capabilities.

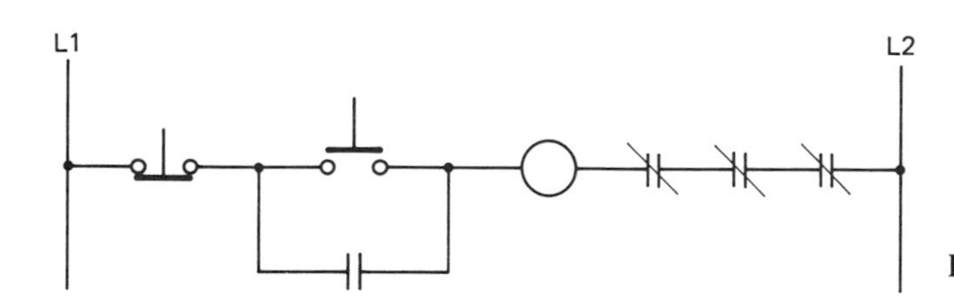

Figure 5-8 Volatile memory circuit.

STATIC CONTROL LOGIC

Problems that usually develop in relay circuits are a result from having moving parts and arcing across the contacts. Static control differs in that there are no moving parts. Transistors are solid-state devices that are incorporated into static control circuits. When used to make decisions for control circuits and packaged in a plug-in plastic case, they are called solid-state or logic modules.

These offer several advantages over relays. The lack of moving parts makes them more reliable, arcing is not present, and the plastic case protects them from the environment. They operate from low-voltage sources which require less power consumption, are smaller, and can be easily changed if defective.

Some disadvantages are their lower tolerance to heat, the need for a separate power source, and the necessity for circuits to raise or lower the operating voltages to match inputs and outputs.

Figure 5-9 shows a typical plug-in-type module. Since the symbols used on these modules vary among the different manufacturers, standard logic symbols that have been adopted for the study of digital logic will be used to represent the different functions of each unit discussed.

ON and OFF Indicators

When discussing ON and OFF conditions in logic circuits, it is helpful to use the number "1" to indicate an ON condition or output, and the number "0" to indicate an OFF or no-output condition. Charts, called truth tables, are used to illustrate all possible input and output combinations. Individual logic circuits are also referred to as gates.

NOT or Inverter Gate

The device used to negate or invert a signal in digital logic is the NOT or inverter gate. Figure 5-10a show its symbol and the corresponding truth table.

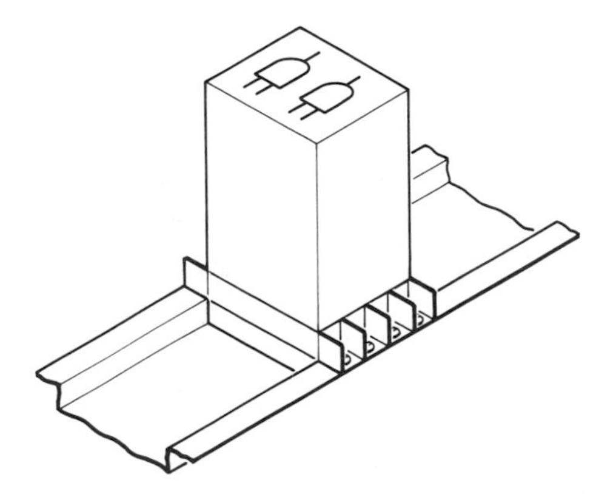

Figure 5-9 Plug-in-type logic module.

A —▷o— B

(a) Logic symbol

A	B
0	1
1	0

(b) Truth table

Figure 5-10 NOT or inverter gate and truth table.

AND Gates

The symbol used for a solid state AND gate is given in Figure 5-11a. This example shows a three-input device. Other variations of inputs are possible, but three inputs are sufficient for this introduction of logic devices. A truth table is shown in Figure 5-11b. By definition, to have an output, all inputs must be present in the AND gate. This can be verified by using the truth table for that circuit.

OR Gate

The OR gate uses the symbol given in Figure 5-12a. The truth table given for the OR gate indicates the condition that if any one of the inputs are present, there will be an output. Note the difference between this symbol and the one used for the AND gate.

NAND Gate

Figure 5-13a shows the symbol for the NAND gate. The NAND gate is really an AND gate with an inverter in the output. Figure 5-14a illustrates an AND gate followed by an inverter and it's truth table. Compare this with the truth table in Figure 5-13b. Recall the conditions that any or all inputs not being present will result in an output.

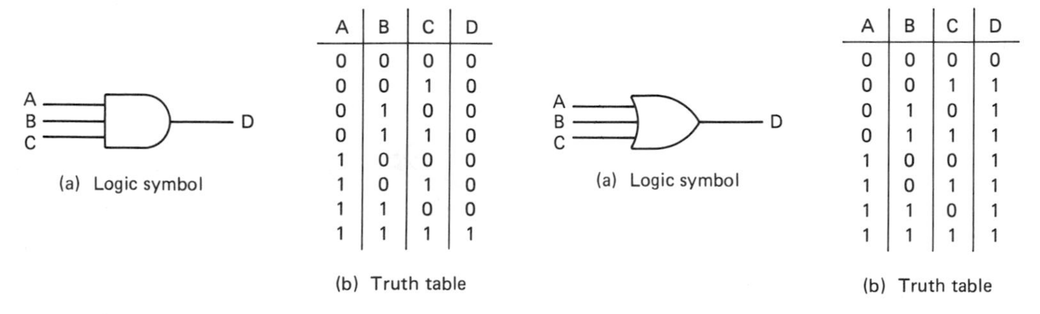

A	B	C	D
0	0	0	0
0	0	1	0
0	1	0	0
0	1	1	0
1	0	0	0
1	0	1	0
1	1	0	0
1	1	1	1

(b) Truth table

Figure 5-11 AND gate and truth table.

A	B	C	D
0	0	0	0
0	0	1	1
0	1	0	1
0	1	1	1
1	0	0	1
1	0	1	1
1	1	0	1
1	1	1	1

(b) Truth table

Figure 5-12 OR gate and truth table.

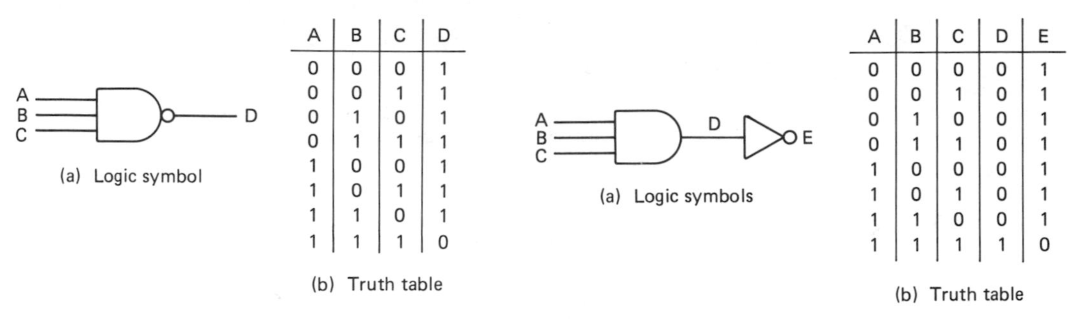

A	B	C	D
0	0	0	1
0	0	1	1
0	1	0	1
0	1	1	1
1	0	0	1
1	0	1	1
1	1	0	1
1	1	1	0

(b) Truth table

Figure 5-13 NAND gate and truth table.

A	B	C	D	E
0	0	0	0	1
0	0	1	0	1
0	1	0	0	1
0	1	1	0	1
1	0	0	0	1
1	0	1	0	1
1	1	0	0	1
1	1	1	1	0

(b) Truth table

Figure 5-14 AND gate with inverter and truth table.

NOR Gate

The NOR gate is illustrated in Figure 5-15a. The specification that an output will be present only when all inputs are not present is shown in the truth table. Figure 5-16 shows that a NOR gate is actually an OR gate followed by an inverter. This is verified in the truth table.

Memory Gates

Two memory gates are possible with solid-state logic—the volatile, or off-return memory, and nonvolatile, or retentive memory. Part (a) of Figure 5-17 shows the symbol and part (b) the truth table for the off-return memory. This gate will lose information stored in it if power is removed from the circuit. Figure 5-18 is the retentive memory gate and its corresponding truth table.

Time-Delay Functions

Time-delay functions are represented by four symbols in industry. In digital logic, a series of counting circuits is used to develop the actual time delay. The

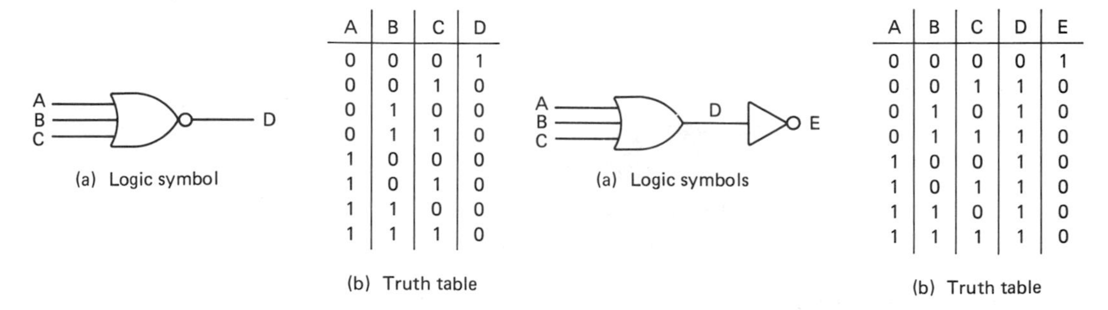

A	B	C	D
0	0	0	1
0	0	1	0
0	1	0	0
0	1	1	0
1	0	0	0
1	0	1	0
1	1	0	0
1	1	1	0

(b) Truth table

Figure 5-15 NOR gate and truth table.

A	B	C	D	E
0	0	0	0	1
0	0	1	1	0
0	1	0	1	0
0	1	1	1	0
1	0	0	1	0
1	0	1	1	0
1	1	0	1	0
1	1	1	1	0

(b) Truth table

Figure 5-16 OR gate with inverter and truth table.

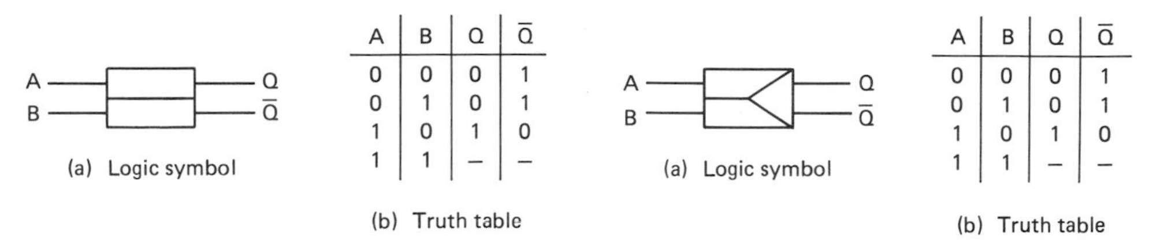

(b) Truth table (Figure 5-17)

A	B	Q	$\bar{Q}$
0	0	0	1
0	1	0	1
1	0	1	0
1	1	–	–

(b) Truth table (Figure 5-18)

A	B	Q	$\bar{Q}$
0	0	0	1
0	1	0	1
1	0	1	0
1	1	–	–

Figure 5-17 Off-return (volatile) memory gate.

Figure 5-18 Retentive (nonvolatile) memory gate.

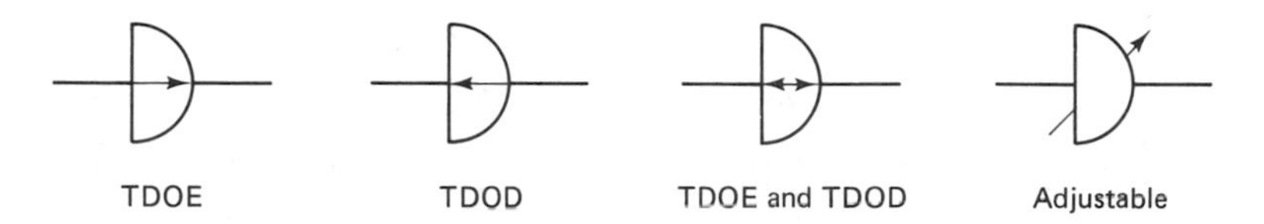

Figure 5-19 Time delay function symbols.

industrial symbols are presented to illustrate these functions. Figure 5-19 shows each symbol, indicating either a time delay on energizing, de-energizing, both, or adjustable.

SUMMARY

This chapter presented both the relay and solid-state logic symbols used in the control industry. Review each circuit and be aware of the conditions that must exist for each function. Solid-state gates will be discussed in the next chapter to illustrate some of the many circuits possible and to give you practice in tracing information in a logic circuit. Variations of some of the gates will also be presented.

EXERCISES

On a separate sheet of paper, complete a response for each question, statement, or problem listed below.

5.1. What is a logic module?

5.2. Give one example of an AND circuit.

5.3. List the criteria for OR circuit operation.

5.4. List the specifications for NAND circuit operation.

5.5. Give one example of a NAND circuit.
5.6. Explain the specifications for NOR circuit operation.
5.7. What does an inverter circuit do?
5.8. Define what is meant by *volatile* or *nonvolatile* memories.
5.9. What is the purpose of a time-delay circuit?
5.10. Identify one problem with the relay circuit.
5.11. List two disadvantages of solid-state modules.
5.12. What is the common name used for all logic circuits?
5.13. How do we represent ON and OFF conditions of a logic gate?
5.14. Draw an inverter and write its truth table.
5.15. Draw the symbol of a 2-input AND gate and write its truth table.
5.16. Draw the symbol of an OR gate and write its truth table.
5.17. Draw the symbol of a 2-input NAND gate and write its truth table.
5.18. In brief terms, what is a NOR gate?

6

Solid-State Logic Circuits

OBJECTIVES

Upon successful completion of this chapter, you should be able to

(1) describe the difference between AC and DC voltage converters, and how they can be utilized with solid-state logic devices
(2) discuss the importance of module power for logic circuits
(3) describe how low-voltage relays and DC amplifiers enable low-level DC signals to control higher level devices
(4) analyze and design basic motor-control or solid-state logic circuits

INTRODUCTION

Solid-state logic components are often used to replace relay circuits. Now that we have described the various functions that can be performed by logic modules we can take a closer look at some of the characteristics that are related to each module and then use them with a practical application.

VOLTAGE CONVERTERS

Logic modules are low-voltage devices that can operate from an AC or DC source. A typical voltage range would be 5V to 15V. Input and output signals can have either a positive or negative polarity, depending on manufacturer design. Not many industrial firms use DC voltages for control of their equipment; this creates a problem if a hybrid system is installed. Remote pilot devices are typically operated at 120 VAC—but this is not compatible with a DC logic module. Because it may be impractical to convert all pilot devices to DC control, the signal input to the logic module is made compatible by the use of a voltage or signal converter. This is a technical way of saying that the signal is rectified and reduced.

The 120-VAC signal is used as the input to a full-wave rectifier, converted to DC voltage, and then applied to a voltage divider network. This method develops a DC signal without creating an additional load on the DC power source used to power individual modules.

Figure 6-1 shows a possible AC to DC conversion circuit with the appropriate logic symbol. You should realize that this circuit does not represent actual component arrangement; it illustrates the *functional arrangement*. The full-wave rectifier could be replaced by a transformer or even a half-wave rectifier. Circuit isolation may also be added by using an optoisolator or relay. Since in most firms repairs are seldom made at this level, signal flow is more important than actual circuit layout within the modules.

DC signals are reduced to operating levels by using voltage divider networks. Figure 6-2 illustrates a DC signal converter with the logic symbol. Isolation techniques may also be used to separate the high-voltage source from the low-voltage output.

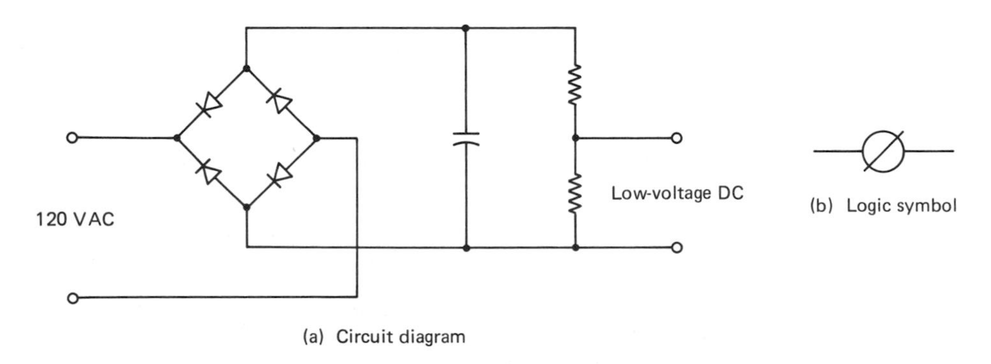

Figure 6-1 AC signal converter.

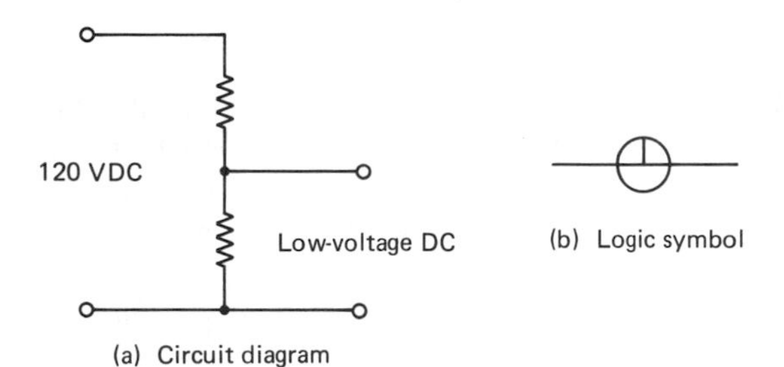

Figure 6-2 DC signal converter.

MODULE POWER

Each logic module requires an active power source in order to function. To simplify electrical drawings, the power connections are usually omitted. Figure 6-3 shows the circuit logic symbol of an AND circuit with the power supply connections included. The actual value of the power source varies with the manufacturer; therefore only a positive and negative DC source is indicated.

DC AMPLIFIER

Low-level DC signals are often used to operate high-level devices. This is best accomplished by using low-voltage relays and an amplifier circuit. Figure 6-4 illustrates the symbol for an amplifier. The internal circuit arrangement varies; but the basic function of all amplifiers is to use a small input signal to control a large output signal. You should compare this symbol with that of the inverter in Figure 5-10a. Notice the small circle in Figure 5-10a from the preceding chapter. That circle represents an inversion or negation of a signal. The inverter is actually an inverting amplifier. A small circle is often used to signify the inversion of an input or an output on a logic gate.

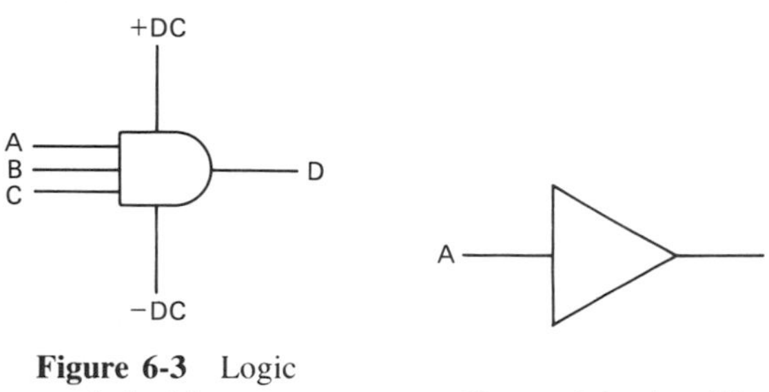

Figure 6-3 Logic symbol with power connections.

Figure 6-4 Amplifier symbol.

Example 1

Figure 6-5 is the solid-state circuit that replaces the relay diagram in Figure 3-3. The requirement still remains that both FS2 and PB1 must be energized before the pump can operate. The signal converters indicate an AC-voltage source for the pilot devices. When a signal is present at both inputs to the AND gate, an output will be produced and the pump coil will be energized.

An easy way of tracing the signal flow through a logic circuit is to assign the symbol "1" when a signal is present and a "0" when a signal is absent. Using the truth tables and assigning possible signal conditions, the circuit may be analyzed. Let's examine the four possible signal patterns with the two-input AND gate.

The first possible situation is that both FS2 and PB1 are open. Figure 6-6 shows the inputs and outputs of this circuit using 0's and 1's.

1. With FS2 open, the input to the AND gate is 0.
2. With PB1 open, the input to the AND gate is 0.
3. The truth table logic for both inputs at zero indicates the output is also 0.
4. A 0 input to the amplifier results in a 0 output.
5. With a 0 output to the relay and pilot light, both will remain de-energized.

If FS2 were to close, a new sequence of events would take place. Figure 6-7 illustrates the conditions of this circuit.

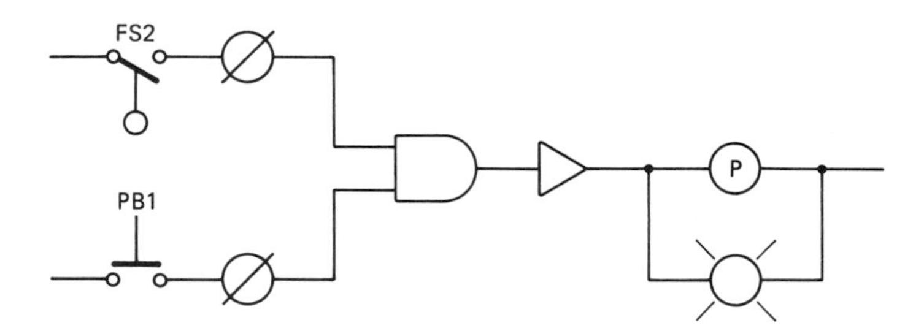

Figure 6-5 Solid-state AND circuit.

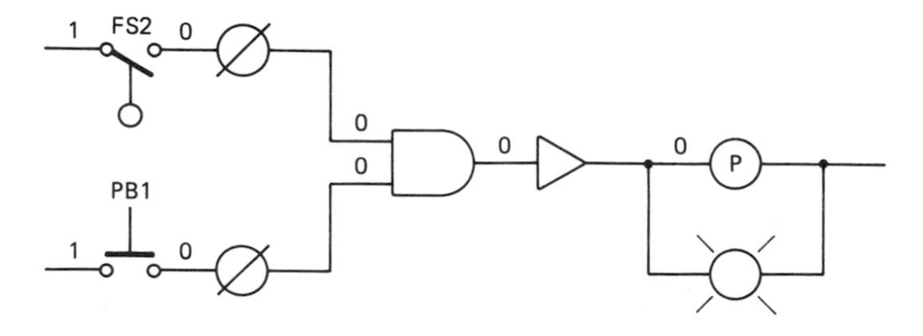

Figure 6-6 FS2 and PB1 open.

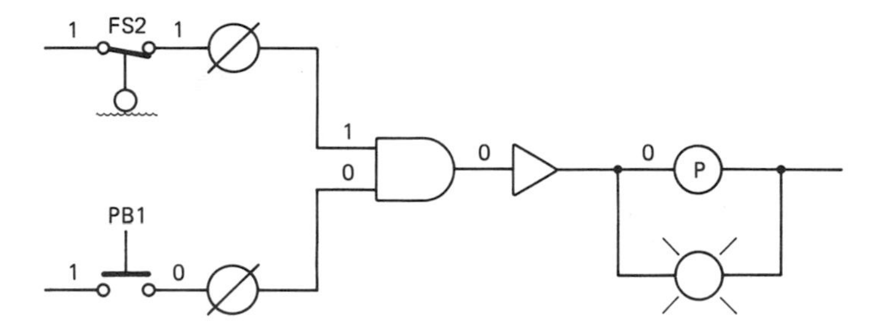

Figure 6-7 FS2 closed and PB1 open.

1. When FS2 closed, a 1 input is present at the logic gate.
2. With PB1 still open, a 0 input is present at the input to the AND gate.
3. The truth table logic for a 1 and a 0 indicates the output should also be a 0.
4. With a 0 output, all remaining components will remain at 0 or a de-energized state.

Now let's assume that FS2 is open and PB1 is closed. These conditions are shown in Figure 6-8.

1. With FS2 open, a 0 input will be present at the AND gate.
2. PB1 closed will result in a 1 input to the AND gate.
3. The truth table logic for a 0 and a 1 will produce a 0 output.
4. A 0 output from the AND gate will result in the relay and pilot remaining de-energized.

In the final condition both FS2 and PB1 are closed. Figure 6-9 shows the signal patterns for this circuit.

1. FS2 closing produces a 1 input to the AND gate.

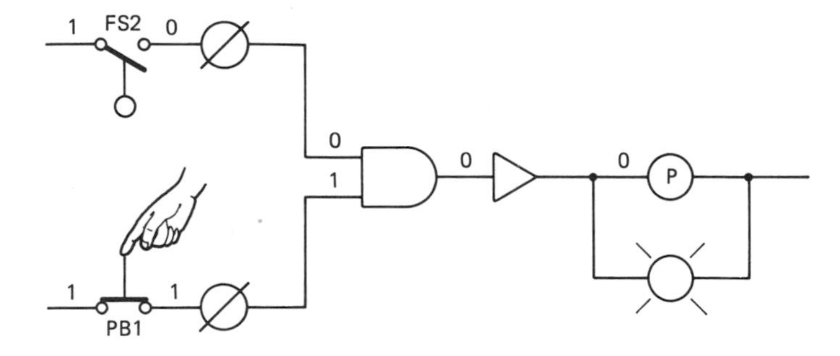

Figure 6-8 FS2 open and PB1 closed.

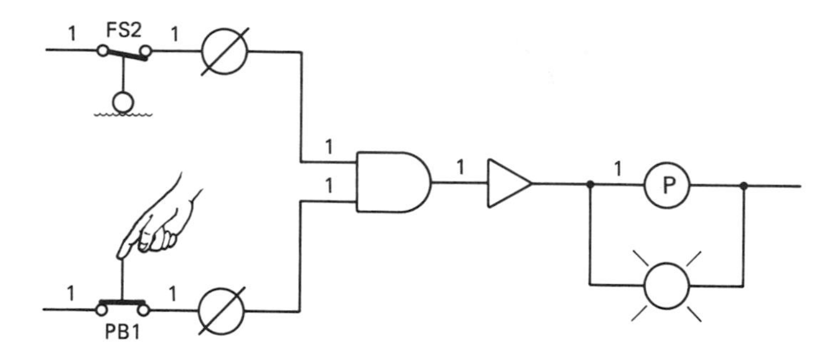

Figure 6-9 FS2 and PB1 closed.

2. PB1 closing also produces a 1 input to the AND gate.
3. The truth table logic with both inputs at 1 indicates that the output will also be a 1.
4. A 1 input to the amplifier produces a 1 output; the relay and pilot light are energized.

By using this method we hope you can see the ease with which a signal can be traced through a logic diagram. The actual magnitude of the signal is not important, since we are tracing signal flow and not analyzing the individual circuit components. Our only concern is to determine if a device is in an ON or an OFF condition.

Example 2

Figure 4-3 is a circuit for a series of conveyors that sequence started. Figure 6-10 is the solid-state equivalent of this circuit. The holding circuit has been replaced by an off-return memory module. Another item that you should notice is that all inputs in this circuit are from momentary-contact, normally-open push buttons. This is due to the characteristic of the off-return memory that requires only momentary inputs to make it function. We will only trace the normal sequence of operations through this circuit. After examining the circuit, you should verify that the specifications originally designed for the relay circuit are still being satisfied. The notation assigned to the push buttons coincides with the same function as in the original relay circuit.

1. To start C3, PB6 is pushed and a logic 1 is present at the input to memory gate, M3.
2. A 1 at the input of M3 will cause it to set (the S in the symbol stands for set) and the output at Q will go to a logic 1 condition. You can verify this with Figure 5-18b.
3. A 1 at Q of M3 will provide a 1 input to AND-gate 2 and also to A3.
4. A 1 at A3 will produce a 1 at the output of A3 and coil C3 will be energized.
5. AND-gate 2 will not produce an output, since only one input to the gate is at a logic 1 state.

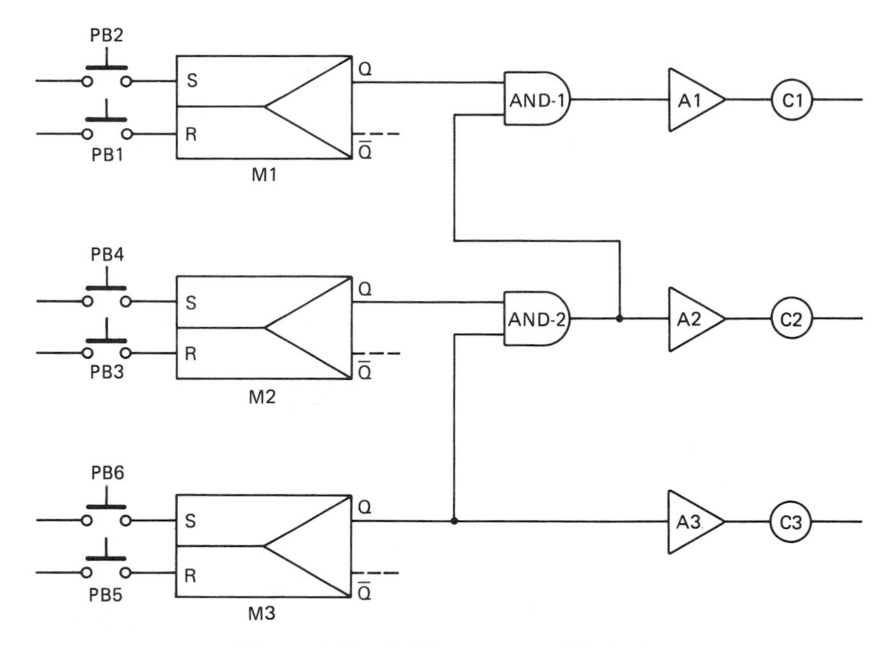

Figure 6-10 Solid-state sequential circuit.

6. Conveyor C3 is now running and all other conveyors are off.
7. Pressing PB4 provides a set signal to M2 and a 1 at its output.
8. With a 1 at Q of M2, AND-gate 2 now has a 1 at both inputs and can produce a 1 at its output.
9. A 1 at the input of A2 results in an output of 1 and coil C2 is now energized.
10. A 1 at the output of AND-gate 2 also provides an input to AND-gate 1.
11. The absence of a signal at the other input to AND-gate 1 keeps its output at 0 and C1 remains off.
12. C3 and C2 are now running and C1 is off.
13. When PB2 is pressed, a set signal is presented to M1 and the output at Q goes to a 1.
14. AND-gate 1 now has both inputs present and sends a 1 to A1, which then energizes C1.
15. All of the conveyors are now running.
16. Pressing PB5 resets M3 (the R in the symbol represents reset) and Q of M3 goes to 0 condition.
17. This causes A3 to go to 0, turning C3 off.
18. Since AND-gate 2 now has one input at 0, its output also goes to 0 and this causes A2 to turn C2 off.
19. AND-gate 1 also has one input at 0, and therefore it causes A1 to turn C1 off.

You should go through the same process of steps 1 through 19; but this time examine the effects of PB3 and PB1. Try doing this before and after all three conveyors are running.

Example 3

You may recall Figure 4-5 in which we added a time-delay sequence to the conveyor system. Figure 6-11 is the solid-state logic equivalent of that circuit. Examine the two circuits and try to visualize the number of moving components present in the relay diagram, and then compare this to the logic diagram. If conventional motor starters are used, you should have counted twelve moving parts in Figure 4-5 and seven in Figure 6-11 (overload contacts were not counted). This may not seem like a great difference to you, but considering that fifty-eight percent of the components with moving parts that may fail due to dust or dirt contamination have been eliminated, it is a significant difference. This is even more dramatic if you can do it to every machine in a factory. The cost of having a system "down" during a process cycle will also determine if the extra price for the logic components will make it economical to replace all of these items.

Two items need to be brought to your attention before we examine the circuit operation. You will notice that each of the AND gates have their inputs in parallel. It is common practice to buy multiple-input gates and connect the unused inputs to an adjacent input. This allows flexibility for future changes and also lets the gate function normally with one input instead of two. When normally-closed switches are used in place of normally-open switches, the NAND gate will provide the proper

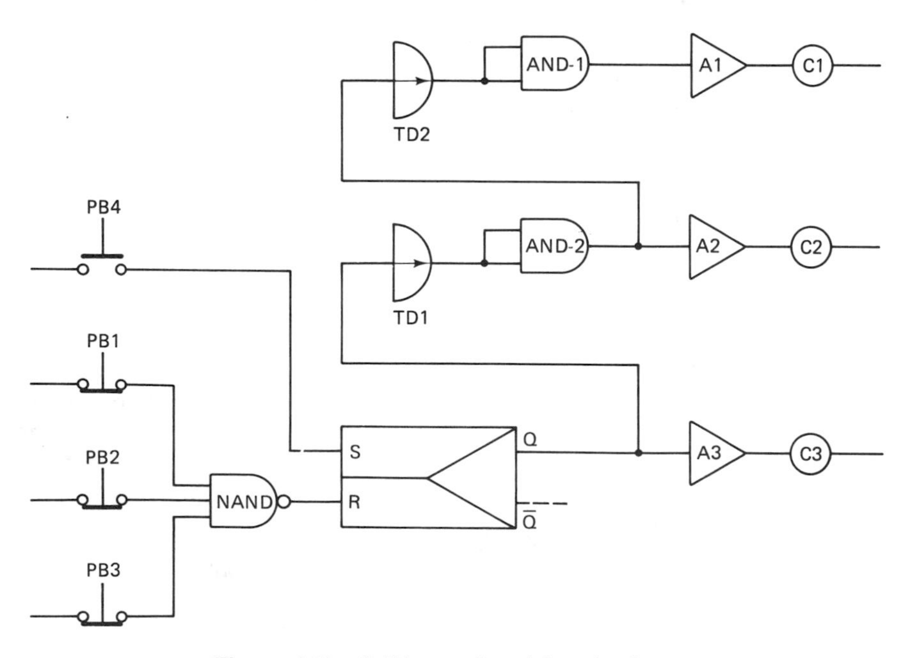

Figure 6-11 Solid-state time-delay circuit.

signals for the specified operation. With this in mind let us go through the step-by-step operation of Figure 6-11.

1. PB1, PB2, and PB3 are all normally-closed. With all three inputs to the NAND gate ON the output will be a 0. This 0 input to M1 has no affect at the present time.
2. If we now push PB4, a 1 input is provided to M1, and this will toggle or switch on output Q to a logic 1 state.
3. A logic 1 to A3 will allow C3 to energize.
4. When Q went to a logic 1 state, the input of TD1 also had a 1 present, and it began its timing cycle. It has a 30-second delay before energizing.
5. After TD1 completes its timing cycle, a 1 is produced at its output and simultaneously to the inputs of AND gate 2.
6. With both inputs of AND gate 2 ON, a logic 1 is present at its output. This output is applied to A2 and TD2.
7. A2 turns on C2 and TD2 begins its timing cycle.
8. After 30 seconds TD2 produces a 1 at its output, and this signal is also present at the inputs to AND gate 1.
9. AND gate 1 will have a 1 at its output and A1 will energize C1.
10. If either PB1, PB2, of PB3 is pushed, the NAND gate will go to a 1 and this will reset M1. Output Q will go to 0.
11. AS Q goes to 0 all other inputs in this circuit will also go to 0 and the three conveyors will all turn off.

SUMMARY

Step-by-step analysis was used again in this chapter to reinforce your knowledge of circuit operations. Hopefully you can see that mentally placing 1's and 0's for signals can simplify analysis of circuit operation considerably. A variety of testers and meters can also be used to measure the actual output of each device. You should now have enough basic knowledge to design and analyze simple motor control and solid-state logic circuits. The ability to develop more complex circuits will require practice and study of existing circuit diagrams. If you feel a lack of confidence at this time, go back and review the material in earlier chapters.

EXERCISES

On a separate sheet of paper, complete a response for each question, statement, or problem listed in the following section.

6.1. What is the typical voltage for a solid-state module?

6.2. What determines the polarity of the input and output signals of a solid-state logic module?

6.3. What is a voltage or current converter?

6.4. Assume that you need a +5-V supply and that your available power source supplies +15 V. How would you go about getting the right supply voltage from what you have available?

6.5. What is an amplifier?

6.6. What does the small circle at the input or output of a logic gate signify?

6.7. When normally-closed switches are used in place of normally-open switches, which of the solid-state logic gates will perform the required operation?

6.8. What is the advantage of using step-by-step analysis of a circuit's operation?

7

Programmable Controllers: An Introduction

OBJECTIVES

Upon successful completion of this chapter, you should be able to

(1) define the term *programmable controller*
(2) describe the fundamental parts of a computer
(3) describe the function of basic computer portions
(4) list several applications of the PC
(5) discuss the basic operation of a PC
(6) select a specific PC for particular applications
(7) explain the meaning of terms or phrases such as *processor*, *computer dead time*, and *operation cycle*.

PC APPLICATIONS

Programmable controllers (PCs) have been going through a development and growth experience much the same as the hand-held calculator. Initially high cost items and limited in function, they are now decreasing in cost and improving in quality. Some of the most common applications are relay replacement in critical short-cycle control circuits. One of the features that makes the PC desirable is the capability of redesigning existing control circuits with little or no wiring changes. This reduces down time and often eliminates installation of new wires, conduits, and components. Troubleshooting control systems is also aided by light-emitting diodes, LEDs, that show the ON or OFF status of an input or output from the controller. Programming terminals and self-test features also enable circuits to be tested without actually applying power to actuate final control elements.

Almost any relay circuit can be replaced with a PC, but it is not always appropriate to do so. It may not be cost effective to buy a PC. They do require added protection from dust and dirt. Clean, filtered air sources for ventilation must often be installed to provide a positive-pressure environment in a control cabinet to keep contaminants out.

BASIC PC OPERATION

You may see little difference between a relay and programmable controller ladder diagram, but they operate on two different principles. The relay circuit operates when all pilot devices have provided a complete current path to the final control element in the circuit. The PC, however, operates similarly to a solid-state logic element. You may recall that control circuits can be divided into information, decision, and final elements. These same elements are located in a PC, and they are responsible for its operation. Let's step through a typical sequence. We shall call this a program cycle.

1. When a PC is first programmed, a data base or a series of preset conditions are stored in the memory of the unit for later use in the form of a program. This section of the computer is examined each time a decision is made by the computer.
2. At a designated time the program cycle is started.
3. Next, the computer examines the information at the inputs.
4. Information is now compared with the data base.
5. The decision section now applies the information to logic-circuit statements designated during programming and determines which outputs receive a logic 1 or 0.
6. All outputs with logic 1's are turned ON and those with 0's remain OFF or are turned OFF.
7. The program cycle is now complete.

8. The computer continues to operate until the program start time is reinitiated and the program cycle is repeated.

This is a very simplified explanation of the internal function of the PC, but it will make it easier to understand the detailed examination of each section that follows. This is also a functional approach that focuses on the user program.

COMPUTER STRUCTURE

All computers are information processing units. They all have a power supply, a timer, a storage area called memory, and a decision area called a processor. The entire system must be operational or the computer will cease to function. Figure 7-1 shows a functional diagram of the basic computer structure.

Power Supply

The power supply provides the necessary voltage or current to each component within the system. If the power supply should fail, erroneous information will be produced or the computer will shut down. For this reason power supplies used in computers are well designed and often have back-up units to provide power in the event of a failure. This is similar to an electrical utility supplying power to your house. If the power fails, your lights will go out, but if you own an emergency generator, you still have power available.

Timing Unit

Another important area is the timing unit. This portion of the system is responsible for providing a common time reference for all the other sections. By synchronizing all the sections, bits of information can be moved within the system at predetermined times or when requested without interfering with the flow of other data that may be moved at the same time. This may be compared to the use of traffic lights and clocks in a town so everyone gets to his or her destination at the correct time. If you were scheduled on an airplane and the traffic lights were not working or you had the incorrect time set in your watch, you might miss your flight. By arriving late, you would have to wait for the next available flight or cancel your trip. The timing section of the computer works in a very similar manner. Information must be present when it is needed, otherwise it may be lost.

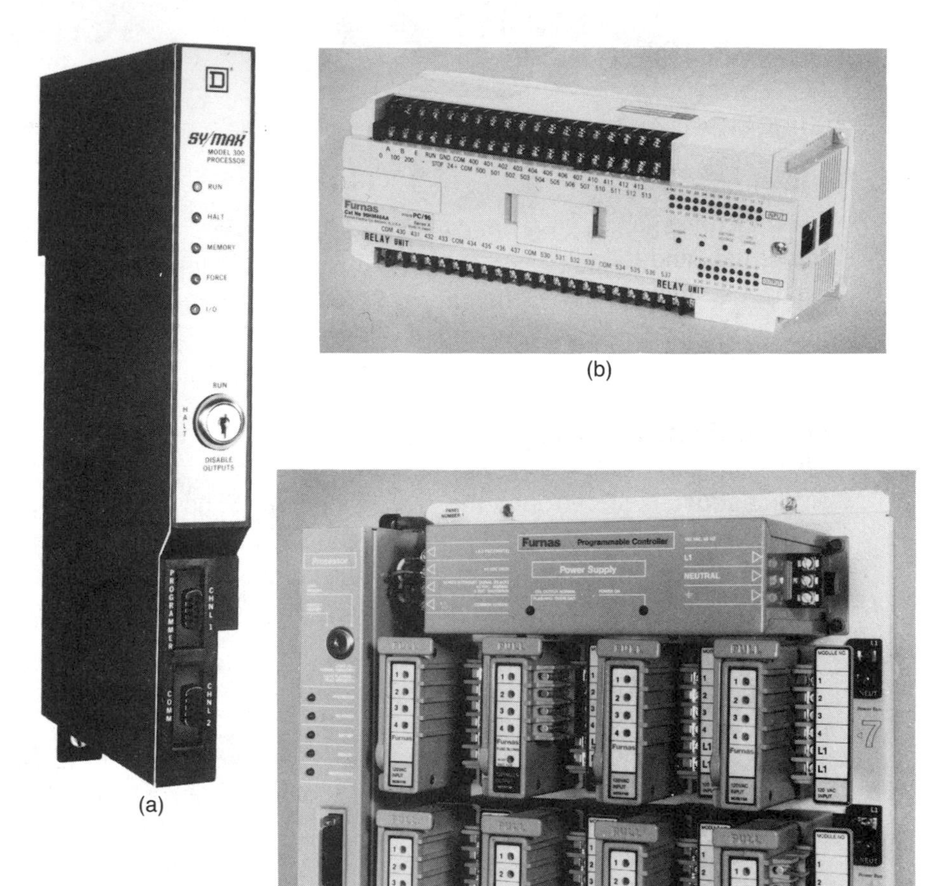

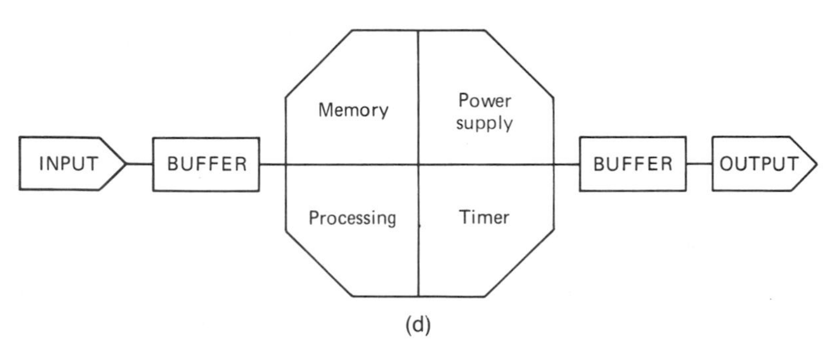

Figure 7-1 (a) Programmable controller processor unit (courtesy of Square D Company). (b) and (c) Programmable controllers (photos courtesy of Furnas Electric Company). (d) Basic computer structure.

Memory

Another important section is the memory. Data is moved in the computer at various times, and it may be combined, compared, or evaluated in some way with other data and then used again at a later time. Suppose you had to remember ten telephone numbers. Knowing you are most efficient doing one thing at a time, you will probably write down each number for use at a later time. Computers, like humans, must store some information for later use, and this is the function of memory. The memory section is usually limited to read, write, or read/write processes only.

Processor

The final section is the processor. This area uses stored information such as mathematical operations, computer operation functions, valid and invalid logic functions, and user-supplied programs to compare data and develop an output based on the data present. An analogy might be your taking a math test. If you know the proper equations and you are given enough information, you can provide a solution for a problem. If you are given invalid data or the problem exceeds your mathematical capability, you will develop an erroneous decision. The computer is vulnerable in the same way. It must operate within the limits of its design, or it will output invalid data if its input data is not valid. The processor does the mathematical calculations, issues its own commands when required, executes commands, and keeps the system organized.

OPERATION CYCLE

The program cycle we examined earlier is only a small portion of the total computer operation. The total *operation cycle* includes all the functions a computer must complete before starting again. Let's examine a hypothetical computer system to see how a typical operation cycle might run.

Figure 7-2 is a diagram of the computer clock cycle—its importance will be more obvious as we complete the discussion of the operation cycle.

As we mentioned earlier, the timing function is very important in a computer. Two MHz may be a little fast for some PC systems, but future systems will surely operate at even higher time bases. A 2 MHz clock means that two million timing pulses are generated in one second. This does not imply that all pulses are used in this time frame, but it gives our system two million separate identifiable timing periods that are available for whatever purpose we choose. Figure 7-3 illustrates a possible computer cycle and its various sections. We will break this

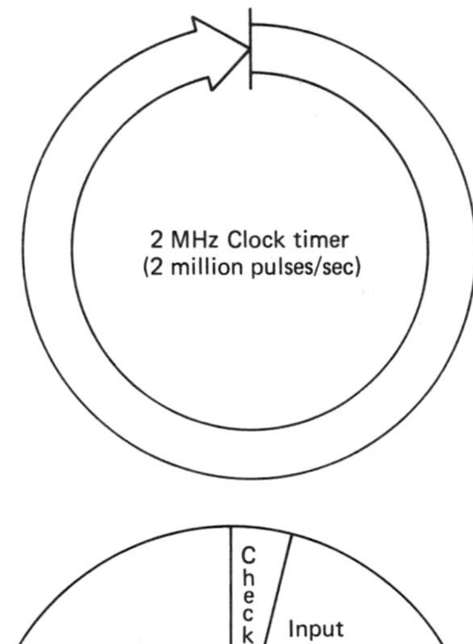

Figure 7-2 Computer clock cycle.

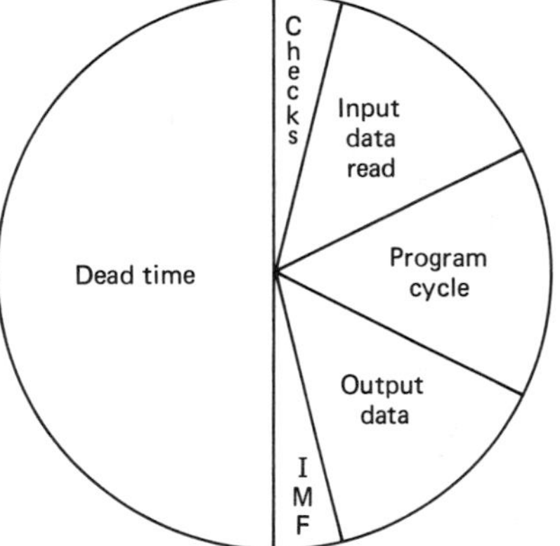

Figure 7-3 Operation cycle.

cycle into individual sections as we discuss them and then relate all the sections to a complete cycle.

1. First, our computer will go through a series of self-checks. We decided when we purchased our system that accuracy was very important. If at any time the computer is not functioning properly, we need to know this. Remember when it was stated that ''faulty information at the input results in faulty information at the output''? In addition to that, if the computer is not capable of producing an accurate output, the results could be very dangerous. Suppose the system just started turning devices on at random. Machinery could be damaged or destroyed, and even worse, someone may be injured or killed as a result of computer outputs. For this reason our system begins each cycle checking various parts of its system to verify that major critical areas are operating properly. Some PCs do an initial check on startup only. If an error is detected, then the system will shut-down, and likewise if everything is OK, the cycle will continue. Figure 7-4 illustrates this part of the cycle.

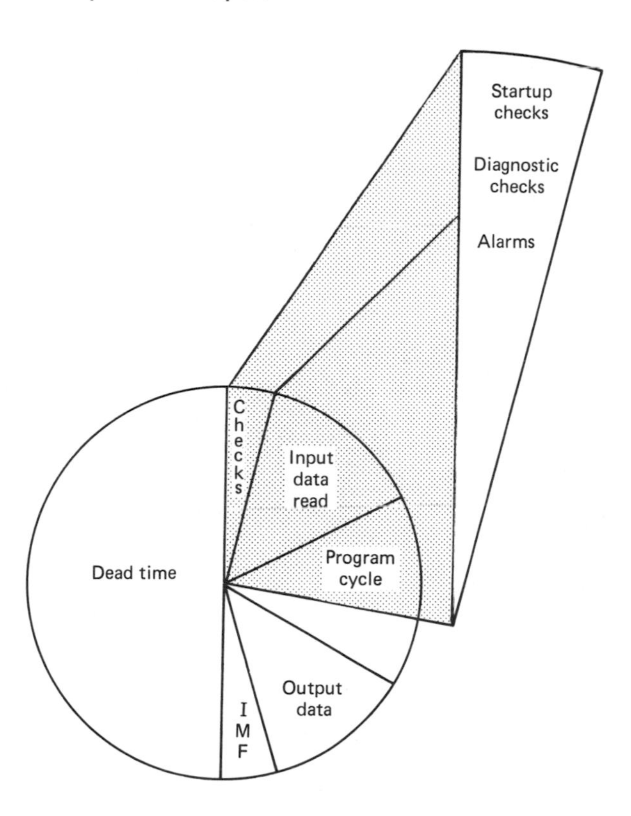

Figure 7-4 Self-check section.

2. Next, data is read at the input. Figure 7-5 shows that this step actually involves several steps. All inputs to a PC are assigned an address or identification code. Each input is read as to its ON or OFF state, and in some cases, analog values may be read as multiple inputs. (Input devices will be examined later as interfacing devices.)

The input signal does not always conform to the computer codes used internally. For this reason each input signal needs to be converted to the appropriate computer code, if necessary.

Many inputs must be sampled, and some method should be used to sort the signal being read from all other inputs. Each section of memory available for an input signal is read at a time designated by the timing signal and the memory is enabled, or turned ON, to allow the information to be stored. When the enable signal is OFF, data can no longer be placed in memory at this location. This process is repeated for each input.

3. The program cycle is now run. This is the part of the cycle which you control. What happens at this time depends on what you store as a program. Examine Figure 7-6 to see how the program fits the operation cycle. Hopefully,

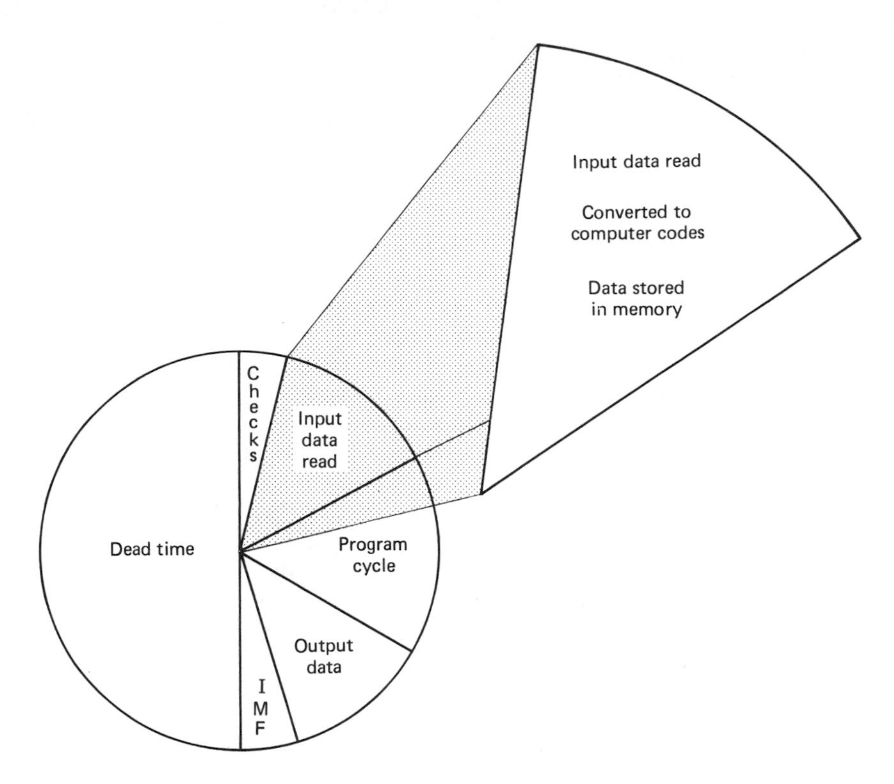

Figure 7-5 Read input data.

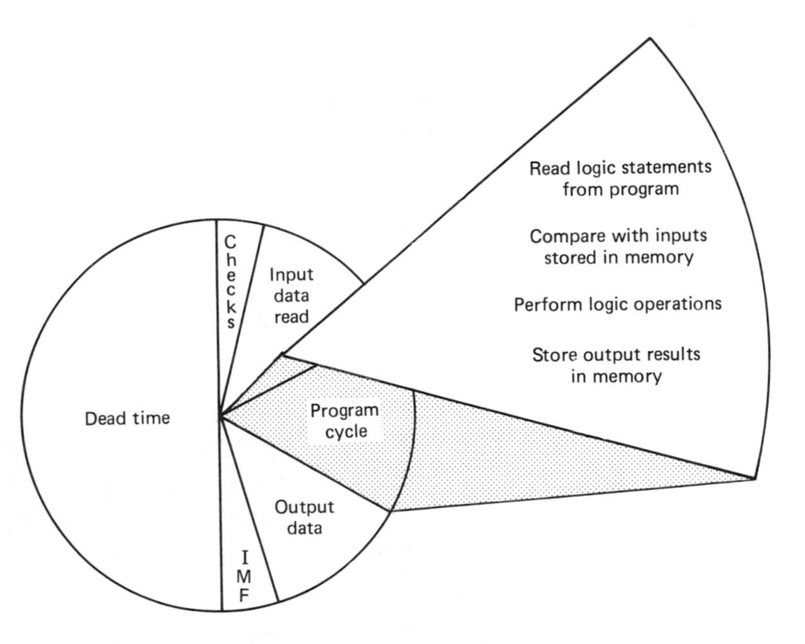

Figure 7-6 Program cycle.

you are beginning to see what a small piece of the pie you have to work with during programming. The programmable space available can be a relatively small area, or it may be expanded by the addition of extra modules. This small space can provide a very valuable function, and its small size should not diminish its importance. (Some of the limits and capabilities of the program will be examined in Chapter 8.)

4. Output data is generated. The program generated a series of instructions for the computer to complete. As each instruction was examined, all the data available was read and compared and a conclusion was developed. Based on the data equation read from the program, an output was generated and stored. Figure 7-7 illustrates how this relates to the operation cycle. If the equation resulted in a logic 1, then the output is turned ON if it was OFF, or remains ON if it was previously ON. A logic 0 will turn an output OFF if it was ON and of course any device previously OFF will remain OFF.

5. Internal management functions. This is time when the computer does a little organizational checking on its self. Figure 7-8 compares this with our operation cycle. It may have to reset a timer, move data, continue a timing

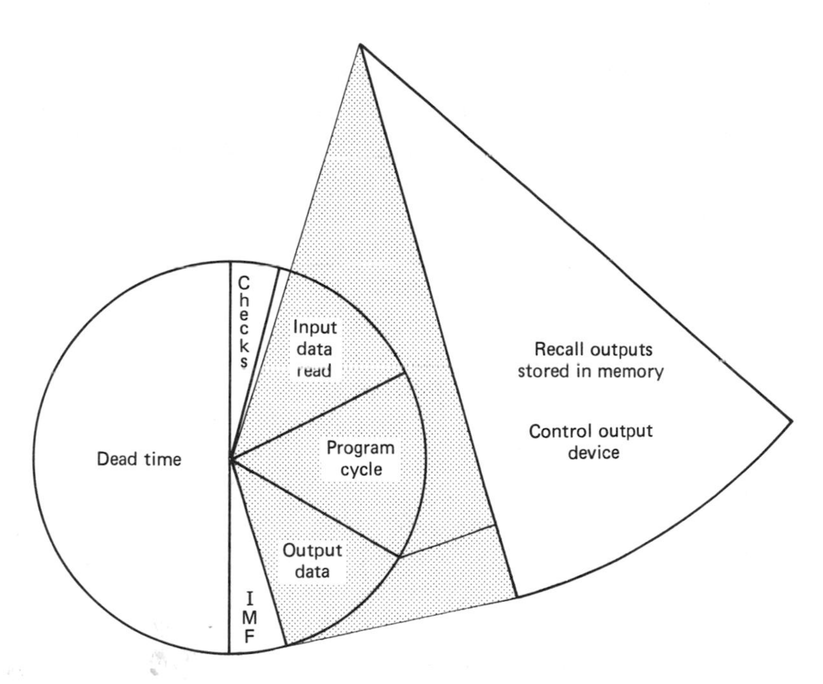

Figure 7-7 Output data generated.

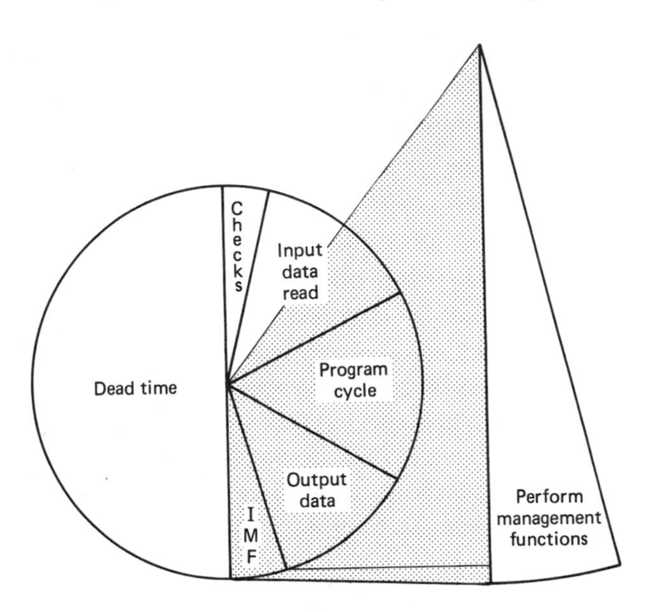

Figure 7-8 Internal management functions.

function, or perform any other related type of process. This is a little like taking time to balance your checkbook to see if everything is in order.

6. Computer dead time. This part of the cycle may or may not exist in your system. If your PC has a fixed amount of space internally for all functions, it will probably be omitted. Economically, it is wasteful, since it reduces the time the computer has to operate. If your PC has a varying program cycle capability, some or all the previous parts of our pie will increase or decrease in size. The dead time in our example is used more to emphasize the importance of time use inside the computer.

Buffers

Computers operate at very high speeds, but data enters into a PC at a relatively slow speed. The buffer is a transition point or a temporary storage area. The computer then periodically samples the input data, while it is not busy with some other operation, translates it to computer codes and then sorts the information, sending it to the correct address within the unit. Since the computer is selecting information to be sent out at a very high speed, a second buffer is used to adjust the output information speed.

The term buffer is also used to describe a circuit internally that protects the computer from improper signals or loads coming into or going out of the system.

Operation Cycle Summary

Now we can put our pie back together and look at the entire operation cycle, compare it to the program cycle, and the clock cycle. Here are the steps of the operation cycle once again.

1. Computer does self-checks.
2. Input data is read.
3. Program cycle is run.
4. Ouput data is generated.
5. Internal Management Functions are Performed.
6. Computer continues timing but performs no functions.

TIMING EXAMPLE

If a chip timer has a clock frequency of 2 MHz and it takes 20 milliseconds to read the input data, process all data, perform all computer management functions, and output data (includes running the program cycle), then there are 40,000 clock pulses used for the entire cycle. This means there are still 1,960,000 clock pulses available for other processor-related functions (the operation cycle requires only 2% of the total clock time in this example). The entire operational cycle can be completed fifty times a second. With another 20 milliseconds dead time between cycles, it is still possible to update and rerun the entire program twenty-five times

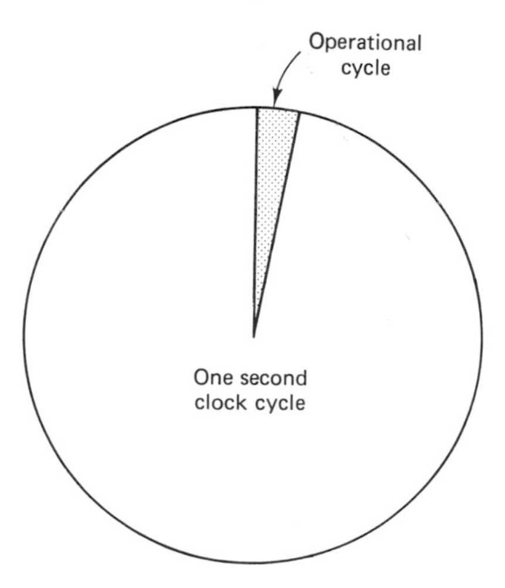

Figure 7-9 Operational vs. clock time.

every second. This is adequate for most control needs. Inputs that require greater accuracy would necessitate increased scanning rates.

If you consider the operational cycle to use 2% of the total time, then we hope you realize the program cycle must use a very small percentage of the total time. Each small section operates at a different time and in sequence. Figure 7-9 shows the time relationship for a one-second time period.

SUMMARY

The purpose of this chapter was to take some of the mystery out of the operation of computers. Our computer was hypothetical and not meant to represent any particular unit on the market. Many features of computer structure and timing functions have not been presented for the sake of simplicity. The organizational structure is representative of what you could encounter and should stimulate your interest to investigate this further. Chapter 8 will describe some typical program structures and requirements.

EXERCISES

On a separate sheet of paper, complete a response for each question, statement, or problem listed below.

7.1. List one common application of the programmable controller.

7.2. The ability to redesign existing control circuits with little or no wiring change makes the PC desirable. What is the advantage of this feature?

7.3. What is the major disadvantage of using PCs to replace relays?

7.4. List four major components of a computer.

7.5. What is the major function of a power supply in a PC?

7.6. What would happen if the power supply would fail to operate?

7.7. What is the major function of a timing unit?

7.8. What would happen when time synchronization between sections is disrupted?

7.9. Describe the basic function of the memory unit.

7.10. The section of the computer whose function is similar to the brain of a human is called ______.

7.11. What is the advantage of the self-check function?

7.12. What happens during the internal management period?

7.13. Discuss the disadvantage of having computer dead time included in our PC.

7.14. Describe the general function of a buffer.

7.15. In the timing example on page 83, if it takes 500 milliseconds to complete the operational cycle, what percentage of the total time is used for this function?

8

Programming Programmable Controllers

OBJECTIVES

Upon successful completion of this chapter, you should be able to

(1) explain the term *programming language*
(2) list and describe PC symbols
(3) explain several procedures for programming a PC
(4) demonstrate procedures and structures for programming
(5) describe the operation and functions of input or output modules.

INTRODUCTION

The intent of this chapter is not to teach you how to program any one programmable controller (PC) in particular, but to provide sufficient information for you to understand the fundamental concepts that are used with most units available on the market today. Upon completion of this chapter, you should consult the operating or programming manual for your particular PC to learn the commands that are unique to your system.

PC SYMBOLS

Since they are nearly identical in appearance, one of the best ways to introduce you to the process of programming is to relate your knowledge of ladder diagrams to the programming format used in most PCs. If you keep in mind that all programmable controllers are computers, you will not be surprised to find computer keyboard symbols used to develop the ladder diagrams. Figure 8-1 shows a comparison of some common PC symbols with their relay equivalent.

Relay symbols are not used for special components such as float switches or push buttons; instead, normally-open or normally-closed contact symbols are used for all pilot devices. All output devices are represented using a pair of parentheses, and special functions are indicated using abbreviations within the particular symbol. Interconnections in the horizontal direction are made with hyphenated lines, exclamation points are used for vertical lines, and a plus sign represents the connection point of a horizontal and vertical line. These are the building blocks used to develop almost all diagrams.

Since all the symbols appear the same, the only way to distinguish one contact from another is by the use of labels. The label appears either above or below each symbol and is unique to an input or output in the controller. By assigning a label, or address as it is commonly called, the location of that operation within the PC is identified. Contacts that are used more than once have

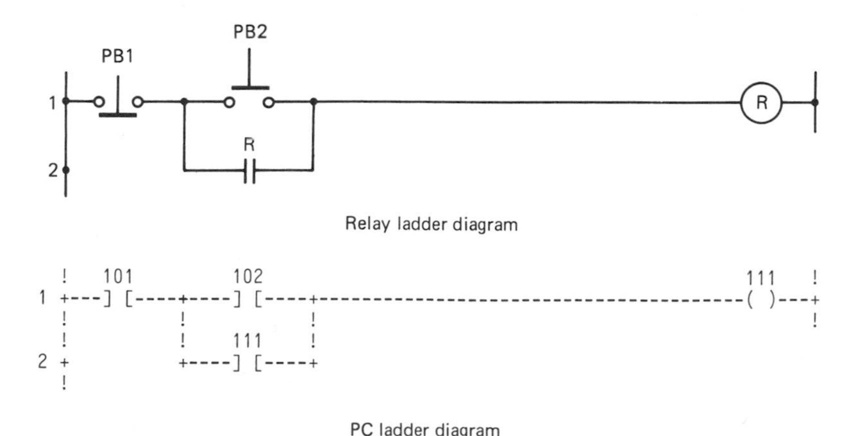

Figure 8-1 Relay and PC ladder diagrams.

the same address. The method of assigning labels may vary with your PC, and your operation manual should be your final reference of all labels and symbols used.

PROGRAMMING

Programming is the process of organizing a sequence of events that we desire the computer to control and then inserting this information in a format the computer can use. The program is read by the computer beginning at step one and then proceeding to each succeeding step as the program may direct. Most PCs do not use written statements like those found on the home computer, although some are available, but instead use the structure of the ladder diagram to step through the program. The computer then reads these diagrams as logic equations or circuits. The transition from relay control to programmable control is best facilitated by using the ladder diagram, and for this reason it will be emphasized.

Methods of Programming

One of the most frequent methods of programming is to use a terminal or programming module. Each has a key pad with the symbols needed to develop a ladder diagram. Terminals may simply have a viewing monitor to see the program structure or may include test features to see the program as it steps through each rung by highlighting it on the monitor screen. Internal diagnostic checks may also be done to examine the operational capability of major components. Programming modules may or may not have displays to show the user inputs. Those that do usually have a limited number of characters on the display at any one time. Both of these are direct manual programming methods.

Programs that are stored and then inserted into memory can also come from other devices. Cassette tapes, floppy disks, and memory chips are often used. Cassette tapes, similar to those found in tape recorders and home computers, store information in the form of 1's and 0's. By selecting the program mode on your PC and running the tape, the program is entered into the PCs memory. Floppy disks perform the same function but are a lot faster than the cassette method. Although many individuals prefer the speed of the floppy disks, the cassette tape is still the most common method. The lack of frequent programming requirements for long periods of time seems to outweigh the benefits of the disk. Disks are susceptible to dust and dirt, and this also contributes to their lack of use. Cassette tapes and their holders have also increased in quality. A good-quality tape insures that all information will be present on the tape as it is programmed and read back by the computer. For this reason they may be rated as *computer quality* tapes by the manufacturer.

Memory chips called erasable programmable read only memories, or EPROMs, are also available. Data stored on these chips, in the form of 1's and 0's can be programmed, stored, and then plugged into the PC. When program changes are made, the EPROM can be erased and a new program installed. Other variations of chip memories are also available. Check your PC owner's manual for the method used by your system.

Types of Memory

Four terms are important when storing your program: random access memory (RAM), read only memory (ROM), volatile, and nonvolatile. RAM memory is usually temporary and is volatile. A power loss will destroy its contents so a back-up source of power is provided, usually in the form of a battery. Since it can easily accept new information, it is used to store programs as well as input and output data. ROM is fixed memory and cannot be changed (except with EPROMs). Programs that are not changed frequently are stored in these chips. Input and output information cannot be stored in this type of chip. The important feature to remember about memory is what will happen if the power to your PC goes off. Will your program be erased? Programs that are lost must be re-entered. It is common for owners of systems that use memory chips to have spare chips programmed just in case the chip in the system is erased or damaged in some way. Programming with this method involves simply replacing the defective chip.

Programming Structure

The first item you must contend with is the physical limit of the programming space. Each PC has a structured area called a matrix. This matrix has a vertical and horizontal dimension that is designed into the computer structure. The horizontal dimension indicates the maximum number of components that may be placed in a rung. The vertical dimension is the total number of rungs available. Figure 8-2 is a diagram of a possible matrix structure. This 10 × 10 matrix can have 100 possible pilot devices. The number of possible outputs allowed and their location depends on the manufacture of the PC.

We will make use of this matrix to program our hypothetical computer. Typical PC symbols will be used. Items that are significant will be noted.

Inputs and Outputs

There is a two-step process to get an input signal into your program and likewise an output signal. Input modules are used to connect pilot devices in the control circuit to the computer. A conventional ladder diagram structure is used. Figure 8-3 shows a possible series of connections to an input module. Each pilot device

```
  !   1    2    3    4    5    6    7    8    9    10   11    !
1 +---] [---] [---] [---] [---] [---] [---] [---] [---] [---] [-+-( )---+
  !                                                            !        !
2 +---] [---] [---] [---] [---] [---] [---] [---] [---] [---] [-+
  !                                                            !
3 +---] [---] [---] [---] [---] [---] [---] [---] [---] [---] [-+
  !                                                            !
4 +---] [---] [---] [---] [---] [---] [---] [---] [---] [---] [-+
  !                                                            !
5 +---] [---] [---] [---] [---] [---] [---] [---] [---] [---] [-+
  !                                                            !
6 +---] [---] [---] [---] [---] [---] [---] [---] [---] [---] [-+
  !                                                            !
7 +---] [---] [---] [---] [---] [---] [---] [---] [---] [---] [-+
  !                                                            !
8 +---] [---] [---] [---] [---] [---] [---] [---] [---] [---] [-+
  !                                                            !
9 +---] [---] [---] [---] [---] [---] [---] [---] [---] [---] [-+
  !                                                            !
10+---] [---] [---] [---] [---] [---] [---] [---] [---] [---] [-+
  !                                                            !
```

Figure 8-2 PC matrix.

has a discrete connection point of its own. PB1 provides information, in the form of voltage L1, to input 101. This unique address will identify the status of PB1 to the computer. Internal circuit operation is not important at this time, but it does represent a load to the pilot device and is connected to L2 to complete the electrical path required for conventional ladder circuits. At this point the input signal is no longer considered an electrical signal but is information stored as a logic value to be used in logic equations within the program. An ON status from a pilot device will be stored as a logic 1. In a similar manner an OFF status will be a logic 0. Figure 8-4 represents the output module. Internal logic values of outputs must be converted back to voltage values in order to operate the final control elements. Power is supplied to the block labeled "110." This represents a control point to turn power on or off to the final control device. (Internal circuit operation will be examined in the chapter dealing with interfacing.)

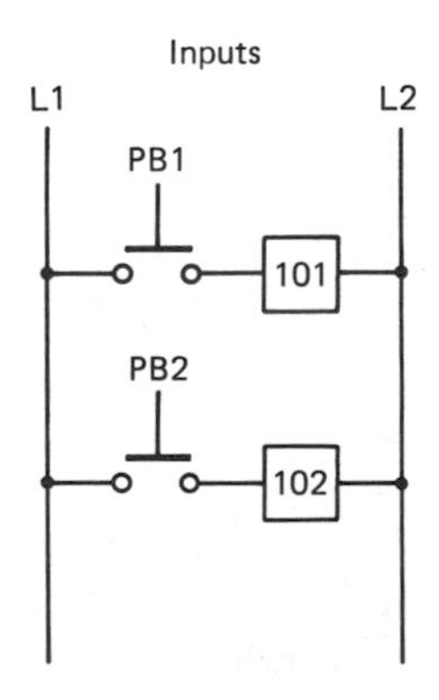

Figure 8-3 Input module connections.

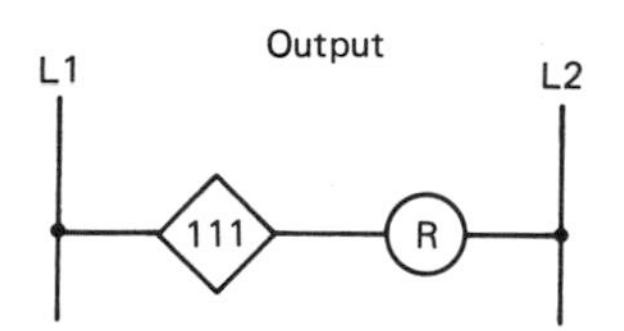

Figure 8-4 Output module connections.

Example 1

The first example is converting a start-stop circuit into PC logic. Figure 8-5 is the complete start-stop circuit from Figure 3-5. Try to remember that we are working with information and logic value, not voltage, in all PC circuits.

1. First of all there are no logic symbols. The status of PB1 and PB2 can be converted to a logic state with the contacts shown.
2. Normally-open *information point* 101 will have a logic 1 or 0 for its value, depending on the status of PB1. Since PB1 is normally-closed, it has a constant ON status until it is pressed.
3. Contacts 102 will normally have a logic 0, since PB2 is a normally-open input.
4. Contacts 110 function similarly to a holding circuit in conventional relay logic. These contacts are tied to the status of output 110.
5. Output 110 is turned ON and OFF by the *logic status* of all the contacts that precede it.

Figure 8-6 is a logic diagram of Figure 8-5. Use your knowledge of start-stop circuits and logic circuits to trace through both diagrams and verify that they are functional.

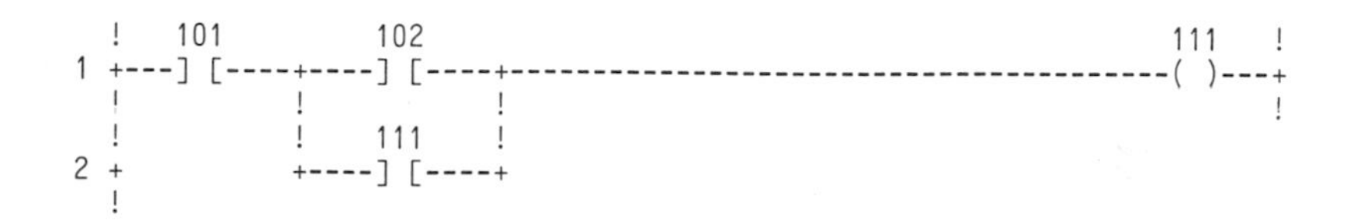

Figure 8-5 PC ladder diagram (Figure 3-5).

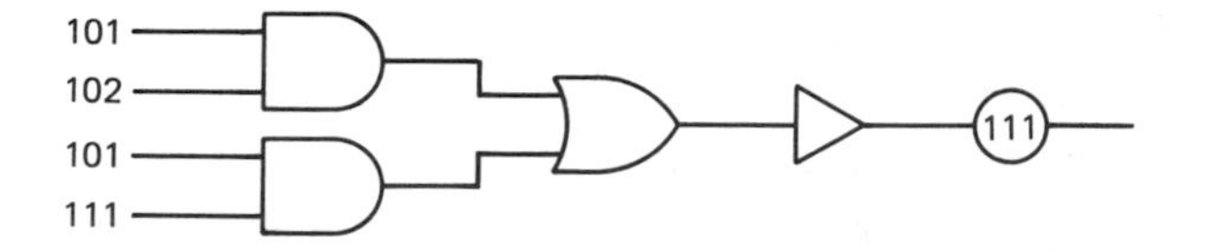

Figure 8-6 Logic diagram equivalent.

Example 2

Our second example takes Figure 3-10 and converts it to a program for a computer. Let's first look at the inputs and outputs required. Figure 8-7 shows PB1, PB2, PB3, and OLs as required inputs and only motor coil M as the output. These are field-wired connections that are brought from the device in the field back to the PC control cabinet. Figure 8-8 shows one possible version of a completed PC ladder diagram.

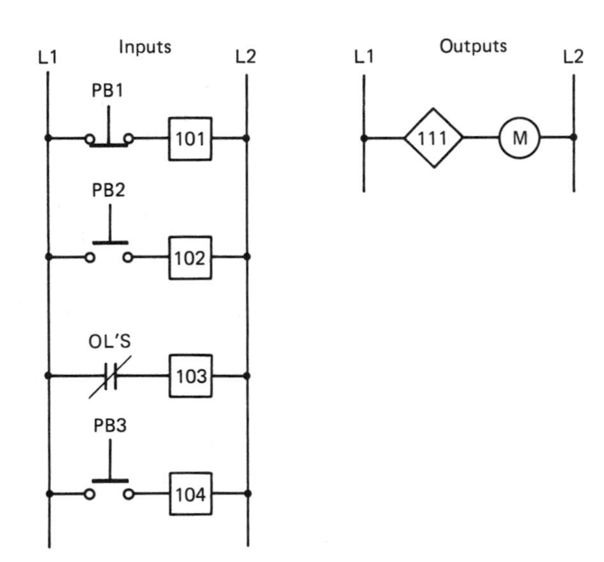

Figure 8-7 Required inputs and outputs.

Let's examine each rung of the program and develop a logic statement for each to verify its validity. You may wish to refer back to Figure 3-10 (Note: PB2 and PB3 will be normally-open contacts on our PC).

1. The relay diagram shows that one possible path for current flow is through the stop button AND, through the start button and then to the coil. It is also a requirement that all current flow through the overload contacts.
2. A logic statement might be If PB1 AND PB2 AND OLs are a logic 1 then M will be ON.
3. PB1 is connected to input 101; therefore the contact used to represent it will have the same label.
4. PB2 is connected to input 102.
5. All three overload contacts can be connected as a single input from field wiring (three relay contacts in series can represent an AND statement).
6. Rung 1 then must have contacts 101, 102, and 103 in series.
7. A second possible path for current flow is through the stop button, through

```
  !  101          102                103                              111  !
1 +---] [---------] [------------+--] [----------------------------( )---+
  !                              !                                        !
  !  101          111     104    !
2 +---] [---------] [----]/[-----+
  !                              !
  !  101          104            !
3 +---] [---------] [------------+
  !
```

Figure 8-8 PC ladder diagram (Figure 3-10).

contact M, through the jog button (normally-open), through the coil and to the overloads.

8. A logic statement for this path would be If PB1 AND contact M AND PB3 and OLs are a logic 1 then M will be ON.
9. Rung 2 then must have contacts 101, 111, 104, and 103 in series.
10. The third possible path is through the stop button, through the jog button, through the coil, then to the overloads.
11. A logic statement might be if PB1 AND PB3 AND OLs are a logic 1 then M will be ON.

Our program in Figure 8-8 contains all three logic statements. Since the overload contacts are a common element in all three, they were put in series in our diagram. The stop button is also a common element and could also be placed in series. Contact 104, in rung two, is shown as a normally-closed contact because of our requirement of PB3 to be normally-open. It also serves as an interlock between rungs two and three. Examine the logic equivalent in Figure 8-9 to see how these three statements can be converted to logic elements. A more efficient version of our circuit is shown in Figure 8-10. The extra contacts used for the stop button have been eliminated.

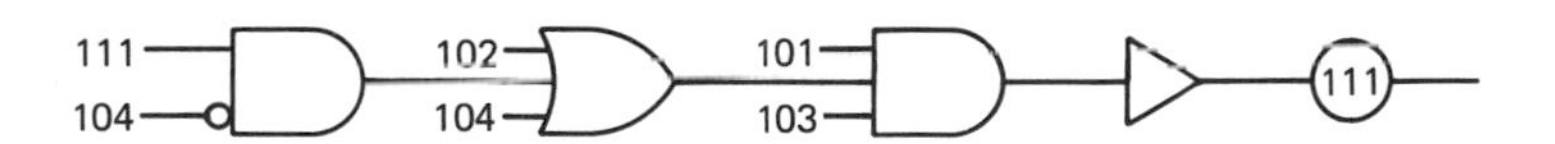

Figure 8-9 Logic equivalent.

```
  !  101         102                103                              111  !
1 +---] [----+----] [------------+--] [----------------------------( )---+
  !          !                   !                                        !
             !   111     104     !
2 +          +----] [----]/[-----+
             !                   !
             !   104             !
3 +          +----] [------------+
```

Figure 10 Modified PC program.

SUMMARY

This chapter introduced you to the basic programming structure that is used by some PCs. Chapter 9 will continue some specific programming structure examples that will allow you to define more complex program statements. If you have a background in Boolean Algebra, you can apply the mathematical equations and binary reduction techniques like Karnaugh maps to minimize the number of logic elements needed in your program. These techniques will not be developed in this book but can be a valuable asset to you if you have the opportunity to learn them. Even without them, you can still write effective programs and by using a little common sense see extra components that can be eliminated.

EXERCISES

On a separate sheet of paper, complete a response for each question, statement, or problem listed below.

8.1. How would you locate an operation within the PC?

8.2. What is the difference between a PC and a computer?

8.3. Explain what is meant by the term *programming*.

8.4. The transition from relay control to programmable control is best facilitated by ______.

8.5. The most common method of programming is to use ______.

8.6. In addition to the memory storage that exists within the computer, list two additional methods of storing information for future use.

8.7. In what forms does the computer store information on cassette tapes?

8.8 In terms of speed, which of the two storage devices is better, cassette tapes or floppy disks?

8.9. For PCs, why is erasable programmable read only memory (EPROM) used more often than programmable read only memory?

8.10. Describe volatile and nonvolatile memories.

8.11. RAM, ROM, and RIW are three versions of memory storage; which one of these cannot be written into?

8.12. How many possible pilot devices can a 10 × 10 matrix have?

9

Additional Programming Circuits

OBJECTIVES

Upon successful completion of this chapter, you should be able to

(1) describe the importance of using input and output diagrams while examining PC programs
(2) discuss the importance of using safety circuits in a PC
(3) describe a procedure for using the master control relay to provide safer operation of a PC
(4) explain the use of timing circuits to accommodate different programming demands
(5) apply MCRs and timing circuits to analyze a conveyor sequence problem
(6) explain the use of latching relays with a PC
(7) for programming PCs, describe the operation and application of up or down counters

INTRODUCTION

Chapter 8 used typical control circuits to introduce you to the programming structure of PCs. The intent of this chapter is not only to convert another relay circuit to a typical PC structure, but in addition, to point out several important features you might encounter in more complex circuits. Safety circuits, master control relays, and timing circuits are among these systems. Figure 9-1 is a sequence-start conveyor system. This relay ladder circuit is to be replaced by the PC control circuit shown in Figure 9-2.

INPUT AND OUTPUT DIAGRAMS

At first glance our program has the general appearance of a relay ladder diagram but the information presented is meaningless without knowing what the address assigned to each represents. Figure 9-3 illustrates inputs and outputs connected to our PC with their devices and addresses. They are written in relay form because

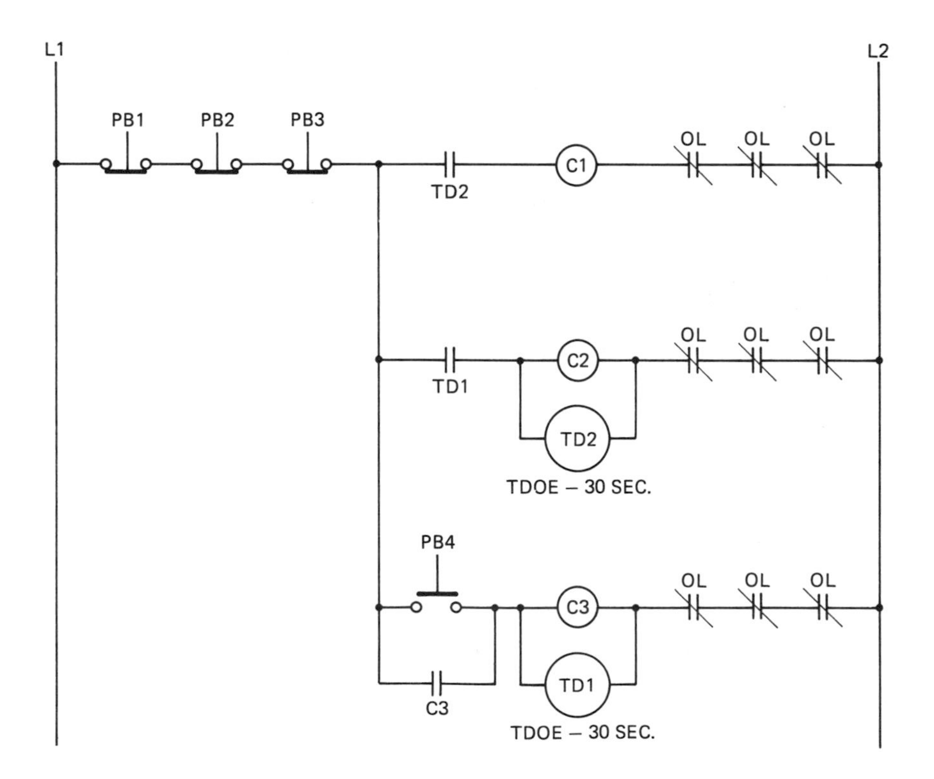

Figure 9-1 Sequence-start conveyor system.

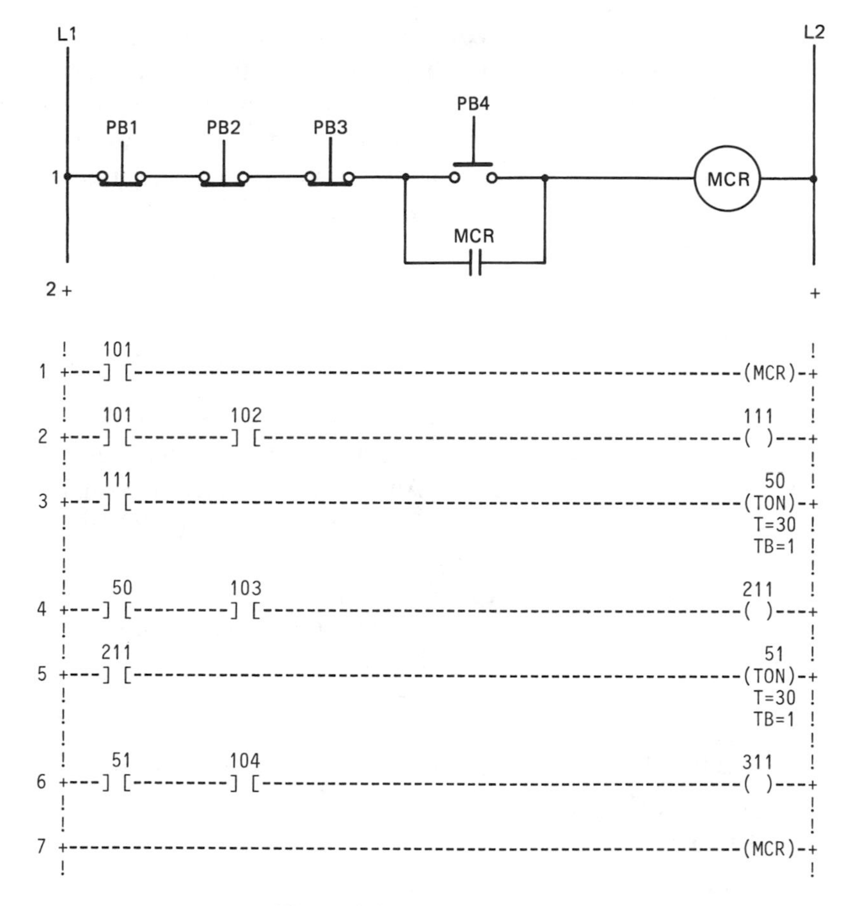

Figure 9-2 PC control circuit.

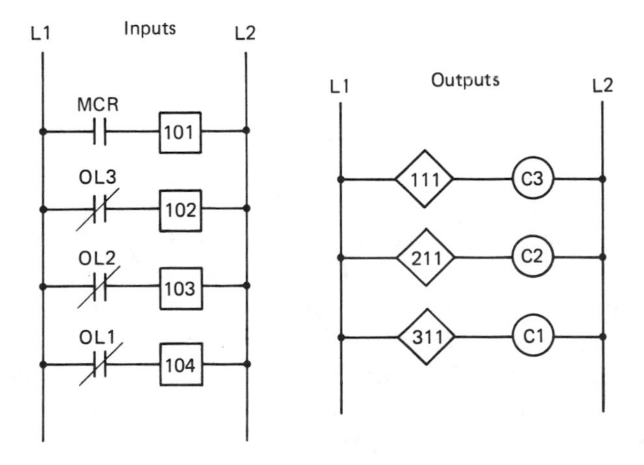

Figure 9-3 Input and output diagram.

they are in fact relay operated. Input pilot devices are drawn using conventional control symbols. The blocks that follow are the actual input connection and address that will be assigned to each particular input signal. The output portion of the diagram uses an output address from the PC as the controlling device, and the final control element is again labeled with conventional symbols. Hopefully you can see the value of having input and output diagrams when you examine the PC program.

SAFETY CIRCUITS

One of the elements of PCs that should be stressed is the need for a safety interlock circuit. The reliability of a PC is very high; but there is also a potential for error, even if it is very slight. Any error, no matter how small, is unacceptable when human life is endangered. Property damage, due to faulty equipment, can be very costly in terms of replacement equipment, down time, and lost production. For these reasons, critical safety-related circuits are left in relay form. One area to consider is what would happen if the circuit suffers a power loss. The solid-lined portion of Figure 9-2 is an example. Let's examine the following sequence of events:

1. PB1, PB2, and PB3 are normally-closed and PB4 is pushed.
2. Relay MCR energizes, closing contacts MCR and providing a holding circuit.
3. MCR remains energized after PB4 is released.
4. Now the circuit experiences a power failure.
5. With no voltage potential from L1 to L2, no current flows through the circuit to energize MCR.
6. MCR is de-energized, and therefore contacts MCR are now open.
7. Now the power is restored and a potential exists from L1 to L2.
8. PB4 and contacts MCR remain open keeping the circuit de-energized.

The simple fact that a relay circuit will not operate if no power is supplied is what makes it so reliable. You need to keep safety a top priority in any circuit you design. If our circuit were allowed to restart when power is restored, it is possible someone might inadvertently get caught in one of the conveyors, material might fall off on someone, or many other potential hazardous conditions might result. In a similar manner, if power was restored to our PC and it experienced a voltage spike that caused a false input, destroyed part of the program, or damaged part of our system, a life-threatening condition could result. Process cycles that must sequence may start in mid-cycle causing equipment to be damaged or destroyed. The safest sequence of events after a power failure still remains to restart the system manually after visually inspecting the process.

MASTER CONTROL RELAYS

An additional method for safety in a circuit is a master control relay (MCR). This relay must be energized before the remaining parts of the circuit can be operational. The degree of safety can be increased by using key locks and other similar devices. The MCR in the relay portion of our diagram is not the same as the MCR in the program portion. Contacts 101 represent the output of the relay portion—this is the transition point from our relay to PC program control.

The output, labeled MCR in our program, controls all sections of the program that appear on rungs two through six. Part of the program structure built into our system stipulates that all rungs that appear after the MCR relay will be operational only if the MCR is energized. If the MCR is not energized, all parts of the circuit that follow it will be de-energized. This would continue through the entire program except that we have allowed the placement of a second MCR output to designate the end of the first master control relay's limits. This allows additional programming in our PC. It should also be pointed out that output MCR is internal to our PC. There are no physical outputs that correspond to it.

TIMING CIRCUITS

Before we examine our PC circuit, we need to address the use of basic timing circuits. Figure 9-4 shows a portion of the program that contains one of the time-delay relays needed for sequencing our conveyors. The address in this case is 50. This will vary with each PC, but in our case timers can have addresses from 50 to 99. Consult the programming manual for your particular unit. The notation "TON" signifies a time on delay and "TOF" is used for a time off delay. Timing is very accurate and can be derived from several different time bases. In our example the time base is one second (TB = 1). This could have been 0.1 s, 1 s, or 0.1 minute, depending upon the capabilities of your PC. Each timing pulse used for the time delay will use the programmed time base and begin counting when energized. This count will continue until a predetermined time, T, is reached.

```
            !
       50   !
----(TON)-!
     T=30   !
     TB=1   !
```

Figure 9-4 Timing circuit.

Figure 9-5 Cascading timers.

Since in our example T = 30, this will correspond to thirty seconds. Although the timer is counting prior to time T, no outputs associated with it will be activated until the timing cycle is complete. The TOF function is identical to the TON function except that all the outputs associated with it will remain on during the timing cycle and go off when it is completed.

Occasionally, timing cycles are not long enough to meet our programming demands. A method to increase the available timing cycle is to cascade or series the timer outputs. Figure 9-5 illustrates this technique. If the maximum count allowed by our computer is 999 and the longest time base is one second, the longest timing cycle available would be 16 minutes and 39 seconds. Let's examine a circuit designed to give us a 19-minute count.

1. Contact 101 closes and timer 50 is allowed to start timing. The remaining portions of our circuit are still de-energized.
2. When the count reaches 999 (16 minutes 39 seconds), all associated outputs will now be energized.
3. Contact 50 now closes, and timer 51 is allowed to start its timing cycle.
4. When the count reaches 141, all outputs controlled by timer 51 will energize (a total count of 999 + 141 = 1140 seconds or 19 minutes).
5. Contact 51 closing corresponds to a 19-minute count.
6. Output 111 is energized after 19 minutes.

CONVEYOR SEQUENCE PROGRAM

We can now examine our program example using your knowledge of MCRs and timing circuits. This example should not be difficult for you to understand. If you do have problems, go back and review some of our previous examples.

1. The MCR from the relay circuit provides an input at point 101.

2. Contact 101 closes and energizes MCR.
3. Since MCR is energized, all remaining rungs prior to the second MCR will be enabled.
4. Contact 101 is enabled with MCR, and if OL3 remains closed, output 111 will be energized starting conveyor three.
5. Contact 111 closes and timer 50 begins its timing cycle. With a time base of one second and a count of 30, the cycle will be completed in 30 seconds.
6. Contact 50 closes at the end of the timing cycle, and if OL2 remains closed, output 211 will be energized starting conveyor two.
7. Contact 211 is now energized and starts timer 51 counting.
8. After 30 seconds contact 51 closes, and if OL1 remains closed, output 311 will be energized and conveyor one can now start.

This circuit operates very much like its relay counterpart. The method is different, but the results are the same. If you still are not sure you understand the information presented, go back and review it now.

LATCHING RELAYS

Several different functions are available with PCs that require special devices in relay circuits. One of these functions is the latching relay. The relay version of this component can be a very expensive item, but many latching relays can be programmed in a PC without any great expense. Figure 9-6 shows the relay and the PC symbols for latching relays. The operation is the same for both relays except the PC relay has the added ability to be latched or unlatched from more than one part of the circuit simply by using the appropriate address.

Figure 9-6 Latching relays.

COUNTING CIRCUITS

Another circuit that is very useful is a counting circuit. Two common versions of the counting circuit are the up counter (CTU) and the down counter (CTD). As their name implies, they count up and down. Additional data must also be

Figure 9-7 Counting circuit.

programmed; this includes the limit of the count. Figure 9-7 shows a short counting relay program with its data. The address of each counter is given above the symbol. The letters inside the symbols designate the purpose of the counter. A reset-counter is designated by the letters CTR. Below each symbol are two values that must be programmed with the counter. First is the present value (PV). This is the count at which we desire some operation to take place. The second is the actual value, or sometimes accumulated value, which is an indication of where the PC has stopped its last count or where we tell the PC to start its count. One important feature about a counter is that it requires only a momentary input. Each pulse will increase or decrease the time of the counter. The input must change states in order to produce another counting impulse. Suppose our circuit is used to control a package sorting and load system. After 30 boxes of our product have passed a certain point, we must activate a gate and push the boxes into a larger box for shipment. Two limit switches, 101 and 102, are used to indicate that a box has passed an inspection point. Let's examine the operation of the circuit with this in mind.

1. When our loading system starts, the PC has set all the values indicated on our program. Our up counter will start at zero and increase until it reaches 30. The down counter will start at 30 and decrease until it reaches zero.
2. The first box comes down the conveyor and pushes both limit switches as it goes past.
3. When the switches close, a logic 1 pulse goes to the counters, and they cycle by one count.
4. Counter 50 had a value of 000, and now it has increased to 001.

5. Counter 51 had a value of 030, and now it has decreased to 029.
6. While the box remains in contact with the switches, no impulse change takes place. After the box is clear, the input signal goes back to a logic 0. This has no affect on the counter.
7. The process continues with each passing box, and each counter continues to change by one.
8. AV = 030 when 30 boxes have passed counter 50.
9. AV = 000 when 30 boxes have passed counter 51.
10. Contacts 50 and 51 now close and send a pulse to the counter reset outputs.
11. The PC now resets the programmed value to the starting point again, and the cycle starts over.
12. A contact with an address of 50 or 51 could have been set to operate a solenoid to close a gate at this time.

One counting circuit would have been adequate for this operation, but two were shown to illustrate the operation of both types. If a readout of each counter were available, it would have been possible to have an indication of the number of boxes that had passed the inspection point and the number remaining until the gate would close. Many circuit variations are possible using counters.

SUMMARY

Your ability to program is dependent upon the knowledge you have of your particular system. The information presented here may overlap several systems, but the basic ideas are still the same. PCs give you a unique opportunity to try your circuits before actually using them on the final control elements. After you have read your manuals, try programming a circuit and running it to see if it is functional. This will avoid the so-called ''smoke test'' that you may have heard of which can be expensive. Chapter 10 will discuss your PC as a unit and will relate several important items to help you understand your system better.

EXERCISES

On a separate sheet of paper, complete a response for each question, statement, or problem listed below.

9.1. What is the advantage of having input and output diagrams?

9.2. In terms of safety, how would you classify the reliability of PCs?

9.3. What would the estimated consequences of a very slight error in PC operation be if the PC were being used to control sophisticated and heavy equipment?

9.4. Why are critical safety-related circuits, in the form of relays, retained in configurations where PCs are used?

9.5. For control circuitry, what is the function of the Master Control Relay (MCR)?

9.6. As mentioned in this text, what is the typical value of the time base?

9.7. What is the major difference between the time on delay, TON, and the time off delay, TOF?

9.8. Why would you cascade timers?

9.9. Two timers are cascaded. Each has a maximum count of 999, which corresponds to 16 minutes, 39 seconds. (a) You desire to start a circuit (energize) after 23 minutes. How many counts are required? (b) How long will the second timer count before the circuit is energized?

9.10. What are two available versions of the counting circuits that are used with relays?

10

Installation Considerations For Programmable Controllers

OBJECTIVES

Upon successful completion of this chapter, you should be able to

(1) list the reasons why installation considerations for programmable controllers are important
(2) list environmental factors that may disrupt the operation of a PC
(3) explain suggestions for preventing environmental disruption of PC operation
(4) identify sources of electromagnetic interference (EMI)
(5) describe the importance of circuit grounding
(6) in different environments, describe how to adequately install a PC

INTRODUCTION

Installing a PC in place of a relay system can eliminate many problems that are associated with hard-wired systems. If the cost of the PC can be justified, you might think your biggest problem is over. This is not the case. The installation of a PC brings with it a set of problems unique to it and other computer-based systems. This chapter will examine some of the major areas of concern with most industrial environments. Try to remember that your particular situation may include many of the items discussed, but not necessarily all of them. There are no clear guidelines that cover all situations, but if you are aware of a specific factor that will affect your operation, be sure to address this when you purchase a PC. The company representative that reviews your particular environment prior to selling you a new unit is serving you well. On the other hand, if a survey of the area to be controlled is not done, you will more than likely have problems later.

ENVIRONMENTAL CONSIDERATIONS

The physical location where the programmable controller is placed is its immediate environment. This is the area of most concern. If this area is considered harsh or corrosive, your happiness with the new system will soon fade as the problems start developing. All is not bleak, however; with a proper enclosure many, if not all, of the major problems can be eliminated. Most enclosures used in an industrial area have a NEMA 12 (National Electrical Manufacturers Association) rating. This is an industrial-rated enclosure that will keep out most dust, dirt, oil, and other similar materials. If the control cabinet must be in a hazardous environment or outdoors, a special enclosure will be required. Since PCs are generally placed indoors, we will consider factors that may occur inside a building.

Any abnormal temperature, excessive amount of moisture, or corrosive atmosphere, for which the computer was not designed, will be considered a harsh environment. A corrosive environment can destroy your system. Three examples of corrosive elements include chemicals, moisture, and temperature.

Chemical

Chemical gases and fumes can react with components and metals on the surface of the printed circuit boards within the PC causing short circuits, false inputs, and resistance changes. The best solution to chemical contaminates is not to place the system in the immediate vicinity of such areas. This is easier said than done, you might be thinking.

Granted you may not be able to move the PC into a special area, but if you first try to place a reasonable amount of distance between the source of the contamination and your PC, you are removing one of many factors that will be a source of difficulties or at least lessening its effects.

Moisture

Moisture is also a corrosive substance. This can be in the form of condensation, rust, or in combination with other chemicals that can create acids. Water that is not in a pure form can be a conductor. If you look at a printed circuit board (PCB), you will notice that the lines etched on it may be spaced very close together. A drop of water could be large enough to physically connect one line to another and complete a circuit. This will surely create problems in the circuitry, since they were separated intentionally. If more droplets accumulate, the problem will compound. Consider what might happen if a resistor had been placed in a circuit to limit the amount of current flowing. If a water droplet short circuited this resistor, a large amount of current might flow in the circuit and destroy a current-sensitive component. Water does not usually get sprayed directly on a PC but is formed as condensation from excessive humidity. There are cases, such as in dairy and food processing facilities, where the area must be washed with large amounts of water. Special care should be taken when near a PC to keep from inadvertently getting it wet. Normally the enclosure will protect it, but when clean-up time arrives and the assembly line is down, maintenance is also done.

In addition to direct moisture on a PCB, water that mixes with some dust on the board can turn into an acid and start dissolving it. This is particularly true around areas with high sulphur concentrations. Hydrogen sulfide is very corrosive and can be destructive in a short period of time. This is true of any area with dust or dirt than can turn to acid with the addition of water.

Temperature

Temperature is not normally corrosive by itself but when combined with water and acids, it can accelerate the corrosion process. Temperature does have a direct affect on some solid-state devices. Increased temperature conditions can shorten component life and increase the potential for failure, which usually occurs at a time that is not very convenient for production or repairs. Temperature effects can be caused by the ambient air around the PC, internal heating of components, or by being placed near a high temperature process. Large industrial firms have a tendency to be either hot in the summer, cold in the winter, or hot all year round. Constant high heat will degrade the components and when combined with internal heat will drastically reduce the life of your PC. Constantly fluctuating temperatures cause the unit to expand and contract, which may lead the PCB to develop cracks in the traces.

Solutions

There are several answers to each of the above problems, and some have common solutions. Distance between the problem and your PC is the first and least expensive answer to most of them. Programmable controllers are affected a lot like people with adverse conditions. If you are too hot standing near an open fire, you will move back. Did you ever notice that the fire will lose more of its effect on you the farther you step back and each step seems to have greater effect than the last step? The same is true with a PC. If you have a direct source of heat near it, every foot back away from the source will have a greater effect of reducing the heat on the unit. This, of course, assumes you are not moving toward another heat source. For just the opposite reason, when you are cold, you will seek a warmer area. The PC functions much better when it operates at its designed temperature. If it is impossible to locate the PC away from all the heat sources, it may be necessary to air condition the control cabinet. This will reduce the temperature but it may also affect the humidity.

If you stand in an area that has noxious fumes, you will again move away to a different location that has clean air to clear your lungs. Since your PC cannot move, we must bring clean air to the unit. Many times a vent is brought directly to the cabinet from an outside or clean air source. You must be careful with this, however. Clean air that is too hot, too cold, or high in humidity will only create a different problem. With a little ingenuity you will usually locate a suitable source. Filters are a must to keep dust and dirt out of a cabinet. The size of the filter will be determined by the size of the dust particles you may encounter. If filters are not kept clean, the air system will be degraded.

We cleaned up the air, removed all the moisture, and cooled the unit down—things ought to be in pretty good shape, right? Well—maybe. If the enclosure is dust proof (and, of course, if we keep the door closed instead of walking away and leaving it open), dust will not be a problem. Even with the best equipment some dust may get into the cabinet. It is a good idea to clean these cabinets periodically with a vacuum to remove any dust or dirt that may accumulate. Dust can coat a PCB and hold the heat of the components on their surface and eventually heat up the entire PCB. This brings us back to the problem of temperature again. It may be necessary to install cooling fans near PCBs that have a tendency to heat up. Usually this is done by the manufacturer of the unit.

Explosive Environment

Generally you would not place a PC in an explosive environment. With a little thought you can avoid this problem entirely. Explosive-proof enclosures are very expensive and must be installed properly to function properly. Just the price of one of these units is enough to get your creative juices flowing to find another way of mounting a PC.

Two more environmental problems are mechanical shock and vibration. Mechanical shock is striking the PC or its cabinet. This could be a barrel bumping against it or a fork truck backing into it. Unfortunately, very little can be done for this other than placing barriers around the cabinet. Vibration is the result of some large force acting on some object near the control cabinet and causing it to vibrate. You can think of it as a miniature earthquake. Not many PCs come with shock mounts, but this is a very effective way of reducing the influence of vibration on the PC. Vibration can eventually cause PCBs to crack and nuts and bolts to loosen. Shock mounts are a small investment compared to the potential damage your unit may experience.

ELECTROMAGNETIC INTERFERENCE

With all the effort we have taken to cure the potential problems with our PC, what else could go wrong? Electromagnetic interference (EMI) is the answer. Radio waves or magnetic waves from a welder can be sources of EMI. They may also come from the wiring system used to bring inputs, outputs, and power to the PC.

Radios

Radios or walki talkis are becoming quite common in factories. If someone would transmit with a radio near your PC, it may induce a false signal into it and create an error in the system. For this reason you should advise anyone with a radio not to operate it near any PC or computer system. Some other control devices are also sensitive to radio waves.

Welders

Welders produce strong magnetic waves and should not be operated near a PC unless the PC is turned off. It's a good idea to reprogram a PC after any shutdown or possible interference from any EMI-producing device.

WIRING

Signal wires that pass in close proximity to power circuits can pick up stray magnetic fields from them and can be sensed as an output from a field device. One of the ways to help reduce this problem is to run all signal wires in steel conduit. The steel acts as a shield and conducts the EMI to ground. In addition to conduit, spacing signal runs away from power runs can also eliminate possible problems. A two-foot spacing will be more than adequate for systems up to 480 VAC. Parallel conduit runs are susceptible to EMI. When you must go near a power run with signal wiring, try to cross it at right angles. Another good preventative measure is never to run power and signal wiring in the same conduit.

Signal wiring to a PC can be of several different types: coaxial or twisted pair cable. Coaxial cable has a single conductor in the center that is insulated from a shield or drain wire on the outside border of the case. Twisted-pair cable has two insulated wires wrapped around each other along with a shield or drain wire. Twisted-pair wiring is more popular because of the ease of running the wire from one location to another and its small size. Each has its own special application depending on the interface used in the system.

Power System Wiring

The power system within the factory can be one of the biggest problems for PCs and computers. Voltage spikes or transients are created every time any device is turned on or off. These spikes can be transferred to signal wiring if appropriate measures are not taken. To clean up the entire system would not be practical, so we take corrective action where it will do the most good—at the PC. All power used by the PC is obtained through an isolation transformer. This transformer does not increase or decrease voltage as most units, but instead has a shielding barrier built in to suppress voltage spikes and noise. It then produces an output equal to the input. This does not mean step-down or step-up isolation transformers are not available.

Noise suppression devices such as thyristors can short-circuit transients around sensitive components and protect them. We will discuss them later when we examine individual components.

POWER SYSTEMS

The power source used for the PC should be given careful consideration. If it is possible to isolate a power source and use it for PCs and computer systems only, many of your problems will be eliminated, especially if you can keep switching mechanisms off the line. These are often called *dedicated lines*. The wiring used for these lines is run in a separate conduit.

The 120-V power source is the most common, but more units are becoming available with either 120- or 220-V power. The higher voltages do use less power, but the PC is not a large power-consuming device. The advantages are not great at this time.

A major consideration may be allowances for power failures with the main power system. The battery backup with the PC will usually be adequate to maintain your program, but it is recommended that the program be re-entered after a system shutdown. This is to insure no false data has been stored as a result of a voltage spike. Extra EPROMs with your program entered can be additional insurance you will not be without the use of your PC for a long period of time after a power failure.

Grounding

Getting a wire from one point to another is not the only thing you must consider. A ground connection must be made somewhere between the PC and its peripheral device. It is *extremely* important that only *one* end of a cable be grounded. This assures there will be no current loops or circulating currents, as they are sometimes called, in the system. A current loop is a difference of potential between two ground points that creates a current between them. This current could be interpreted as an input by the PC, or it could interfere with normal signal flow. Improper grounding will definitely cause you to have headaches from your system. Avoid them early by using proper grounding procedures.

SUMMARY

We hope you are not discouraged by this chapter—thinking that PCs are nothing but trouble. It is only fair that you know the negative sides of a system as well as the advantages. Programmable controllers do need some extra consideration if they are to remain trouble free. The benefits derived from proper installation will not only eliminate your PC problems but will help develop good wiring procedures for relay circuits as well. The next chapter will address interfacing and what to look for or specify when buying a system.

EXERCISES

On a separate sheet of paper, complete a response for each question, statement, or problem listed below.

10.1. What does the term *immediate environment* mean?

10.2. What feature of the PC would eliminate most of the major environmental problems?

10.3. What does the term *NEMA 12* stand for? Why is it most commonly used for industrial environments?

10.4. List the characteristics of a *harsh* environment.

10.5. Identify three elements of a corrosive atmosphere.

10.6. What is a first and least expensive solution to most environmental problems with a PC?

10.7. What is the best protection for an explosive environment?

10.8. Mechanical shock and vibration are two more ______ problems.

10.9. Identify common sources of electromagnetic interference.

10.10. How might a radio affect the operation of your PC?

10.11. Explain one of the ways to reduce the effects of EMI.

10.12. How can we protect a PC from line-voltage spikes?

10.13. Can an isolation transformer be used to step up or step down voltages?

10.14. Name another device that suppresses noise.

10.15. To avoid ______, it is important to have only one end of a cable grounded.

11

Interfacing

OBJECTIVES

Upon successful completion of this chapter, you should be able to

(1) define interfacing
(2) describe how to interface components with a personal computer
(3) list and describe various types of interfacing
(4) describe what is meant by *output overlaoding*, *leakage current*, or *drive capability*
(5) describe the *network arrangements* and *configurations* interfacing systems
(6) read and interpret interface standards for MAP, RS-232-C, and RS-422

INTRODUCTION

This chapter will introduce you to a very important aspect of controls—interfacing. Many times when you purchase new equipment, you have great aspirations for all the new functions it can perform. This thrill can soon fade away when you realize you have bought only a basic functional unit and additional equipment must be purchased to do all the neat things you expected it to perform. Interfacing allows one functional unit, with its own format of operation, to communicate with another functional unit that has a different format of operation.

INTERFACING EXAMPLE

The term *interfacing* is used in many different situations. We will examine the electrical aspect of the term which also has many separate meanings. Let's look at a simple example of how important interfacing can be and then take a closer look at some important interfacing circuits.

Home Computer Example

Many of you may have purchased a personal computer for use at home. Interfacing is a very important feature that should have been considered during the purchase. Assuming you want to have a top-notch set-up, you would buy the basic microprocessor, a monitor and a printer. The good honest salesperson will not only describe all the functions your computer is capable of performing but will also explain its limitations. After a brief little demonstration you may think this is great and just what you need, so you tell the salesperson you will take it. You purchase your system, take it home, and then spend hours learning how to operate it. Perhaps you realize it would sure be nice if you could print graphics on the printer. After a quick call to your local computer store, you are told your system does have this capability, but you must purchase an additional printed circuit board. This new PCB is usually called a *printer interface*. The purpose of this interface is to take information that has been generated in the computer in one format and convert it to a format the printer will understand. Since the keyboard entry that occurs when you type a letter goes into the computer as an ASCII (American Standard Code for Information Interchange) code, which represents a single letter or character, and the printer uses a dot matrix method to print letters, two different formats are being used. Figure 11-1 illustrates how the letter "A" might appear in a dot matrix format in comparison to a typed character. The dot matrix printer requires some means of translating an ASCII code into a dot matrix code. In this example the interface is simply a decoder. The information originally entered represented the letter "A", and the final character is the letter "A"; the data has been transferred from the keyboard to paper with the same meaning but in a completely different format. In this example a mechanical input, the

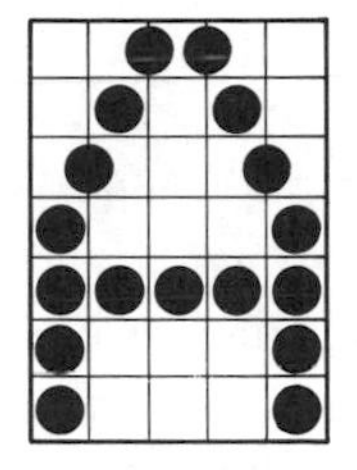

Figure 11-1 Interfaced output.

keypad, was converted into an electrical signal. This signal was translated into a different electrical code, and then converted back to a mechanical output, the print head of the printer. That's interfacing.

TYPES OF INTERFACING

Although many people associate the word interfacing with computers, this is not entirely correct. Interfacing may involve converting from a mechanical signal to an electrical signal, an electrical signal to a mechanical signal, an electrical signal to another type of electrical signal, or a mechanical signal to another type of mechanical signal. A specific device used to transform a signal in one form of energy to another is called a transducer. A transducer can convert into any of the formats mentioned, but usually it is either mechanical-electrical or electrical-mechanical. These will be discussed later as sensors. They are mentioned here to expose you to the wide range of devices that involve interfacing and to make you aware of other possible uses for the term.

Computer Interfaces

Programmable controllers and microprocessors require some type of interface to communicate with the controls they receive information from or that they operate. These can be categorized into four main groups: AC inputs, DC inputs, AC outputs, and DC outputs. The variations in the type of input or output allow additional flexibility in the control environment. This flexibility permits a PC to be manufactured and operated from the many different control voltages and signals that can be found even in the same factory. The internal functions of PCs typically operate from DC voltages. The interfaces discussed will examine methods of getting input signals to appropriate levels for the PC to use them and will convert PC signals to levels that can operate the final control elements in the field. These are normally built into the PC, and you need only to attach the lead wires to the terminal connections.

AC Input Interface

Pilot devices used for controls switch voltages normally in the 120-volt range, since this is a common control voltage. Let's assume we have a push button that is an input to a PC. When the contacts close, the 120-volt potential is present at the PC and, in this case, represents an ON condition. The 120-volt AC voltage is too high for the PC to use, so we must use an AC input interface to reduce the voltage to a lower magnitude and convert it to a DC potential. The diagram in Figure 11-2 illustrates this circuit. The first part of the circuit is a rectifier and a series resistor. The rectifier is used to change the AC input voltage to a DC voltage. The series resistor functions to reduce the input potential to a lower value. Although we have accomplished two of the necessary steps to convert our input signal to a level the computer can use, the circuit still does not apply this new input directly to the input portion of the PC. Located between the rectifier and the computer is an optoisolator and a series of circuits known as filters and buffers.

The optoisolator is a very important part of the circuit. It serves as a protection or isolation barrier between the high voltages outside the computer and the lower voltages inside the computer. Its name comes from two terms *optics* and *isolation*, which mean light isolation. This name describes how the optoisolator operates. There are two components, the light-emitting diode and a light-sensitive transistor, located in a sealed unit. When an input is presented at the PC, current flows from the rectifier circuit and through the current-limiting resistor to the LED. The LED produces light, which can be either visible or infrared, and the lens on the LED focuses this on the light-sensitive transistor. The transistor acts as a switch because it is either ON or OFF when it is exposed to light from the diode. The switching operation cycles the output to an ON or OFF condition at the filter and buffer level, which then goes to the input of the computer. This is illustrated

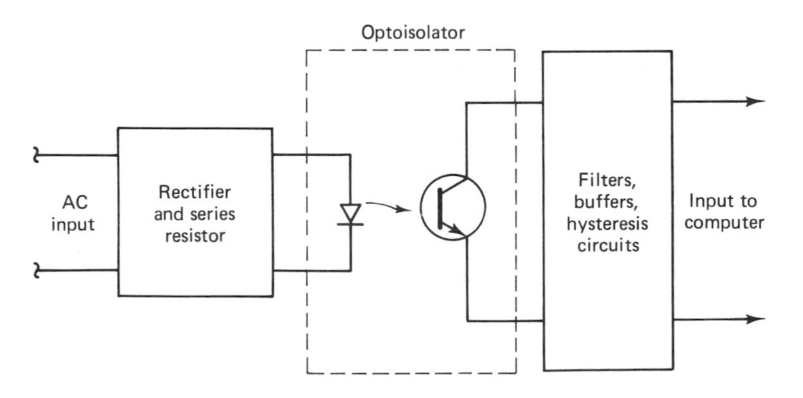

Figure 11-2 AC input interface.

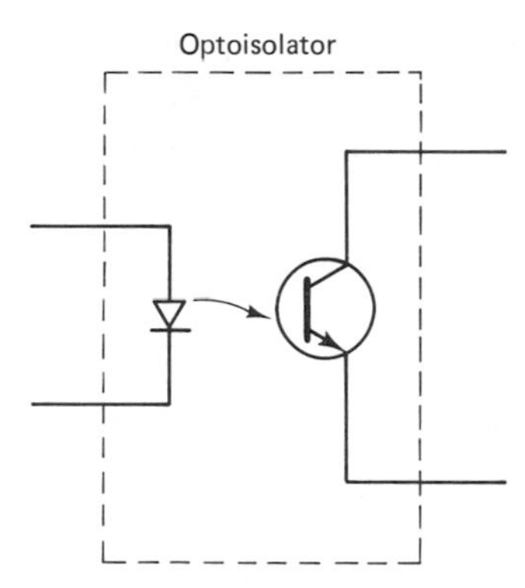

Figure 11-3 Optoisolator operation.

in Figure 11-3. Actual operation of the rectifier, the filter, and buffer stages will be discussed in a later chapter. The important thing to understand here is the functional operation of the circuit, not the individual components.

This may seem like a lot of extra components to use just to get a signal to the computer, but let's see how it isolates and protects. The actual optoisolator has only two components (this is simplified for discussion) that have been discussed so far. A third, but very important part of the circuit, is the air gap between the diode and the transistor. This air gap serves to keep the input to the optoisolator electrically separated from the output. Two conditions must be met to have current flow: a potential difference and a complete path for current. Across the air gap we have a potential difference present but not a complete path for current to flow through. At the low potential of 120 volts, this air gap offers a very high resistance, which prevents the potential from jumping or arcing across it, but at the same time offers little or no resistance to the light waves that can transverse it. This is similar to some remote control TVs that use infrared light to change channels and volume. The remote unit has a battery and a circuit to operate the LED inside it (which represents a potential difference change 0). The space between you and your TV is the air gap. The receiving unit in the TV is the light-sensitive transistor and its associated circuitry. The greater the distance between the diode and transistor in each case increases the resistance of the air gap. If we could guarantee that the potential at the diode would not exceed a certain level, the 120-volt isolation would be adequate. Remember that, because of the rectifier and series resistor, the actual voltage on the diode is considerably less than 120 volts. We are not so fortunate in industrial environments. When voltage potentials are switched ON and OFF, transients or voltage spikes are created that may be many times greater than our normal operation voltage. For this reason, optoisolators are designed to have isolation protection values, determined by the air gap, in the range of four kilovolts to seven kilovolts. This is a small investment to protect your controller from damage and is standard on nearly all PCs. In addition, many optoisolators also provided transient voltage protection by means of resistors and capacitors that form snubbing circuits. These circuits dissipate the transient conditions and protect sensitive components from high voltage potentials that may exist in the interface, under both normal and adverse conditions. The rated degree of protection specified, of course, assumes you are operating the optoisolator within the manufacturer's specifications.

It may appear that there are three separate modules in the AC input interface. Normally all of the circuits are built into a single module that is interchangeable with the other interfacing circuits whose function is identified by a color code on the outside of the unit.

DC Input Interface

A DC voltage is a second type of possible input. The DC input interface has the same purpose as the AC input interface, except it does not have to convert the input voltage from AC to DC. A voltage divider network replaces the rectifier and lowers the input potential to a usable level for driving the LED ON or OFF. Figure 11-4 shows this in functional form. The operations of the optoisolator, filter and buffer circuit are identical to those found in the AC input interface. When your needs change, and DC input signals are replacing AC input signals, the only requirement to update your system is the replacement of the input module.

AC Output Interface

Individual circuits within a PC do not require high voltages or currents to function. This is one of the features which makes it possible to add many different types of circuits to the computer system as a whole unit and not demand a large power source. These small operating voltages are not very helpful, however, when we need to operate some control element outside the computer. The computer does not operate any device directly but instead provides an indication to the output interface about what logic statement needs to be presented. It is the function of the output interface to supply the necessary signal and power to operate a particular device.

Figure 11-5 shows a typical AC output interface circuit. The low-level power from the computer is used to illuminate the LED of the optoisolator. The low-magnitude requirements help prevent loading down the computer and degrading its performance. Isolation protection is once again used to keep harmful transients in the output from damaging the computer. Once the light-sensitive transistor turns ON or OFF, it allows another special circuit, called a triac circuit, to switch a higher AC voltage ON or OFF and then drive the AC load. (Triacs will

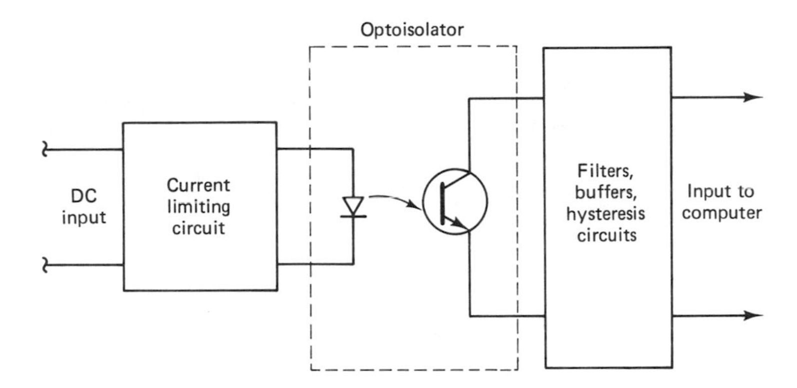

Figure 11-4 DC input interface.

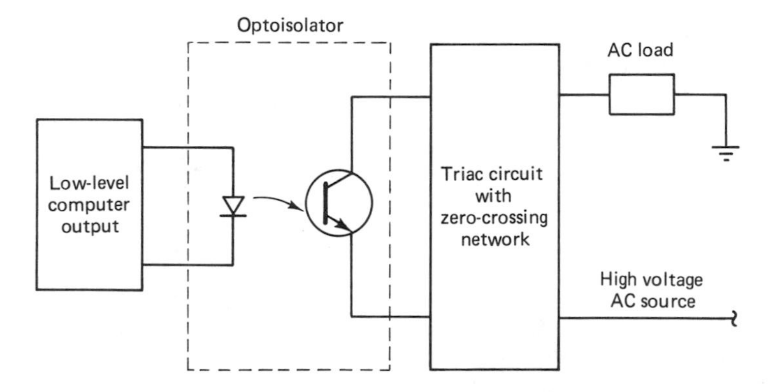

Figure 11-5 AC output interface.

be discussed in a later chapter.) The high AC voltage power that is used in the output is provided by a source separate from the interface unit.

DC Output Interface

Occasionally, DC voltages are required to operate final control elements from a PC. The DC output interface serves this purpose. The operation is similar to the AC output interface except that a DC amplifier or switching network replaces the triac circuit. Again it is not the function of the PC to supply the operating power to drive circuit components; this is done by a separate DC power source. Figure 11-6 shows a diagram of a typical DC output interface.

Converting from types of output circuits can be easily done by replacing the existing module with the new output desired. To match the new module, you may also need to replace the power source.

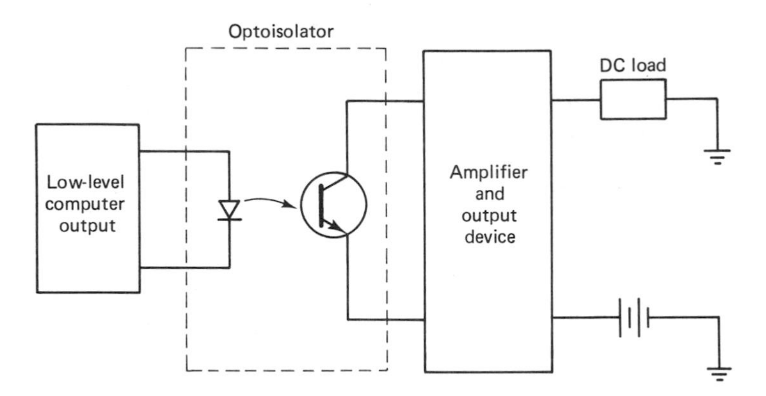

Figure 11-6 DC output interface.

OUTPUT LOADING

Circuit loading is one of the most overlooked items when final control elements are connected to interface devices. Each interface element has a rated output voltage and current. Many times these ratings are not considered, and too many components are connected to a single output interface. This causes the interface to malfunction. When it is necessary to use one output to control many functions, you should consider driving a low power relay with contact ratings to match your load requirements. This allows the interface unit to supply power at its rated output and load conditions and still meet your need for additional outputs. This arrangement, shown in Figure 11-7, serves as a mechanical amplifier and still satisfies our definition of an interface circuit.

Transient voltages, generated by inductive components, can also overload an interface circuit. If this occurs, a relay circuit will usually take care of the problem.

LEAKAGE CURRENT

A problem associated with many solid-state devices is leakage current. Interfaces that contain solid-state relays and triacs may have a small current that flows even when the component is considered to be in an OFF condition. This can be hazardous if it would inadvertently turn on an output device. When this condition is present, a relay which requires a greater operating voltage may be the answer. Never let the additional cost of an item or the trouble to install it interfere with the safety of personnel or equipment.

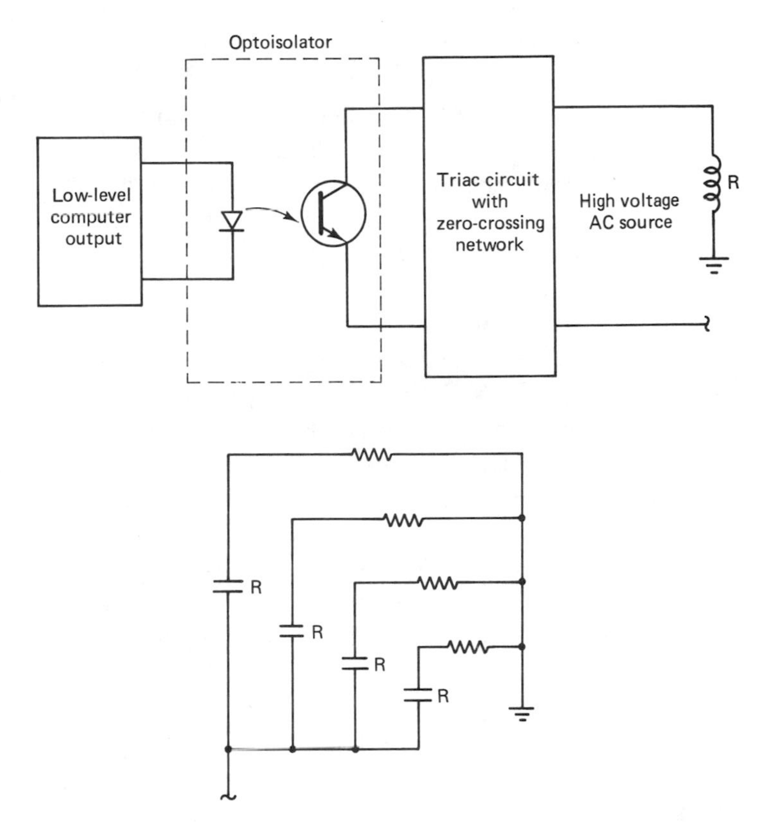

Figure 11-7 Mechanical amplifier and interface.

INTERFACING SYSTEMS

The systems discussed so far have been at the PC or microprocessor level. It is important to meet the requirements at this level, but eventually, as systems are upgraded, the interfacing problem will grow from the individual unit to a system or network level. A network may have as few as two computers connected to it, or many computers. The main purpose of a network system is to allow either a centralized point of control at a main computer or communication between several computers. Capabilities of networking systems are expanding faster than they are being implemented at the factory level. Only a few major industrial organizations have presently installed or tested the newest systems, since standards have yet to be fully accepted. A brief description of the different types of network arrangements that are used, and some of the standards that are involved, is given to provide you with an insight to future systems you may encounter.

Network Arrangements

A network is very similar to a telephone system: in fact many of the interconnections are completed by telephone or land lines. Consider a two-way radio with a microphone to transmit and a receiver to listen. When you push the mike button, you are not able to receive a signal. Some computer systems operate the same way when they communicate with each other. Only one computer can be on the line at a time. This is called a half-duplex link. This may be adequate for some operations, but as systems improve in capability, full-duplex operation will become a more accepted standard. Full duplex allows both units to communicate with each other at the same time. Strictly speaking, it only ''appears'' that things are taking place at the same time. Because of the speed at which computers operate, the time is shared between both functions and what is actually going on is not apparent to us humans. This is more like our home telephone: we can talk and listen to someone at the same time, but we run the risk of losing part of our communication. If we alternately talk and listen, our communication is better organized, and we do not lose information. This is also what is being done with computers. Figure 11-8 shows both of these systems.

Two further categories of transmission format are *asynchronous* and *synchronous*. Asynchronous transmissions are free to transmit at any time between bits of data, while synchronous transmissions are sent during a specified time period. This is a common mode, since it is more organized and therefore more economical.

Network Configurations

Our basic discussion up to this time has been with point-to-point systems. A more flexible arrangement involves one main computer, the host, and many separate units called satellites. This is the star connection shown in Figure 11-9.

In this system the host computer will interrogate each individual satellite unit to see if there is information to be transmitted. If the satellite responds

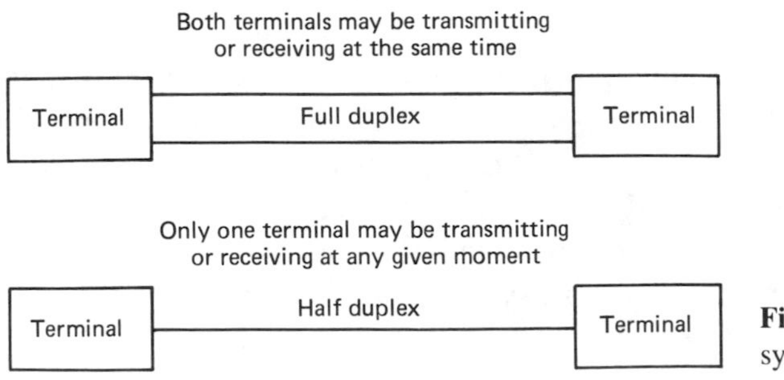

Figure 11-8 Full- and half-duplex systems.

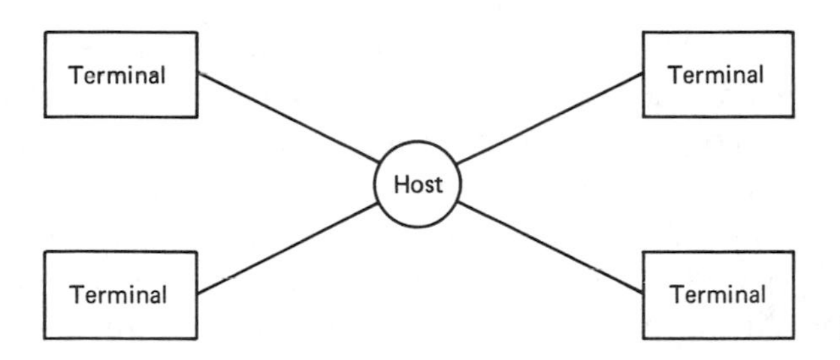

Figure 11-9 Star network.

"yes," then the host gives an "OK" signal and communications begin. Depending on the configuration, communication can continue for a definite time period or until all information has been received. This sequenced interrogation is called polling. This system has the disadvantage of being very slow as far as computers are concerned, and it gets even slower as more satellite units are added. The host must poll each satellite before information is transmitted. As more units are added, the time between polling increases and thus the response time decreases. The advantage of this system is that the host has access to each satellite at any time.

An equally important network is when no host computer exists but each satellite can alternate as a host or satellite unit. This is called a ring or hybrid ring. Figure 11-10 illustrates both systems. The basic ring arrangement has no host computer but sends an electrical signal called a *token* to each satellite unit. When the satellite unit possesses the token, it has the right to act as a host computer and

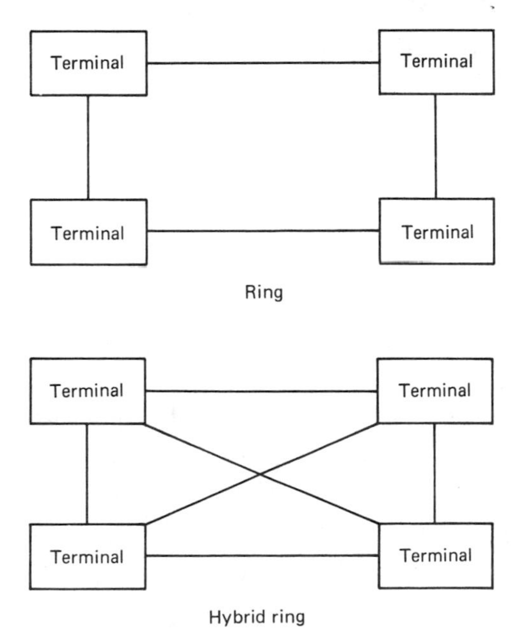

Figure 11-10 Ring and hybrid ring networks.

transmit data to the entire network. After a predetermined time the token is passed to the next unit to act as the host, and data is again transmitted. The time that the host can possess the token can be of a definite time length, a length determined by the amount of data to transmit, or no time if the unit has no data to send. The advantage of this arrangement is that data is constantly being processed and units with no data to send are by-passed, thus reducing the OFF time of the system. The disadvantage is that a break in one part of the ring may disable the entire network. This is overcome by the hybrid ring, which provides multiple paths for the token.

A final example is the multidrop network. Figure 11-11 shows this network. In this configuration there is usually a host computer, and each system is addressed when data is transmitted or received. As long as no other signal is on the line, information is passed back and forth between host and satellite. The advantage of this network is that no one satellite can deter the integrity of the entire system. If a PC, or satellite, was removed from one process and relocated to another, the network would continue to operate. It is also less expensive to run a single wire from one end of a plant to another. However, there is always the chance that a break in the line will disconnect part of the system.

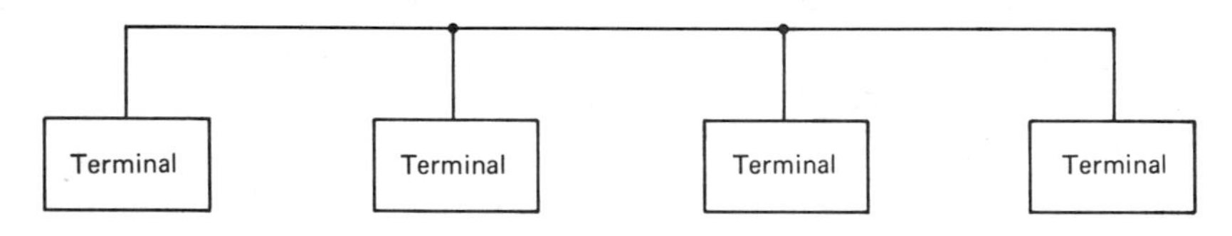

Figure 11-11 Multidrop network.

STANDARDS

A common point of reference for the transmission of data is called a standard. Standards are usually an indication of the success of some individual product. After a product has met the needs of the market, similar products are produced by many different firms to help satisfy the demand—but all use the same basic format. Once the need for a standard is determined, representatives of interested groups meet with standard-forming organizations, such as the Institute of Electrical and Electronic Engineers (IEEE) and the Electronic Industries Association (EIA) and formally establish the standard. The standard includes not only the format of the data but rules called *protocols*. These rules then serve as guidelines for new products and connections to existing systems to help make more systems compatible.

RS-232-C

Timing cycles have been divided into transmit and receive times to provide a time for both to alternate sending data out and receiving it. Sometimes the receiving unit must give the "go ahead, send the message, I'm ready" signal to the other system before it can transmit. This is commonly called *Clear to send* (CTS). Following a transmission, the sending unit may indicate "that's all of my message, you're clear to transmit." In simpler terms this means a *Request to send* (RTS). In addition to the signals mentioned, both units have to indicate that they are ready. These signals form part of a standard called RS-232-C. This standard is used for data processing information transfer and is serial in nature. That is, the information is sent out in a series of single pulses, one after the other. When speed is not essential and the information does not have to travel a long distance, this is an acceptable standard. Not all of the required signals of the RS-232-C standard were discussed. Some of the other signals are used to check the integrity of the system as well as timing functions. This standard does not adapt well to the long distances needed within a factory. It does not function well past fifty feet and does have frequency limitations that are unacceptable for industrial use.

RS-422

This standard uses a balanced system of wiring and has good operating characteristics up to 4000 feet and at high frequencies. As the distance of transmissions are decreased, the effective level of frequency operation increases. In addition to its long-range operation it also has good noise-rejection characteristics. This is important in the industrial environment, since noise levels may often exceed normal signal levels, resulting in a loss of information. The long transmission line does have the disadvantage of introducing line losses, which affect not only the amplitude of the signal but also the loading or impedance which the signal generator must transmit into.

MAP System

The General Motors Manufacturing Automation Protocol System (MAP) is the standard most likely to be used in future industrial environments. This system, which is in a multidrop format, is designed with a manufacturing environment in mind. Central control of all automation and programmable controllers will eventually be possible with this system. One of the key features is that the use of many different PCs is possible, since each protocol is interfaced onto a central data highway. The interface is two way, allowing transfer of information among two dissimilar systems. What is most significant about this is that it has the support of nearly all major computer and programmable controller manufacturers.

SUMMARY

This chapter examined some of the highlights of interfacing. Many items were not discussed. We hope you realize how simple one interface can be, and on the other hand, how another can be very complex. Not all factories will use protocols such as the MAP, but as new systems are developed, the availability of increased control over entire automation systems will make them worth considering.

Chapter 12 will look at what is important in specifying new equipment and can serve as a guide in what to look for in design specifications. It will also help you determine areas that may need improvement in existing equipment.

EXERCISES

On a separate sheet of paper, complete a response for each question, statement, or problem listed below.

11.1. What is interfacing?

11.2. What does ASCII stand for?

11.3. Name the two major types of interfacing.

11.4. List the most common types of inputs and outputs to and from the PC or its microprocessor.

11.5. What typical voltages do the PC internal functions operate on?

11.6. What determines the isolation protection values?

11.7. What is a typical range for the isolation protection values?

11.8. What would be used to replace the rectifier in the AC input when a DC input is applied?

11.9. Describe the function of an output interface.

11.10. What is the purpose of isolation protection?

11.11. In the AC output interface, a triac is used to switch a high AC voltage; what replaces the triac for a DC output interface?

11.12. Describe the consequence of overloading an output interface unit.

11.13. Explain the main purpose of the interfacing system or network.

11.14. What is a half-duplex link? What is a full-duplex link?

11.15. List the two categories of transmission format.

11.16. Name the network configuration in which the host computer controls many separate units.

11.17. What is the disadvantage of the star configuration?

11.18. List the two major components of a standard.

11.19. Describe the three major interface standards that are being used in the industry.

12

Design Specifications

OBJECTIVES

Upon successful completion of this chapter, you should be able to

(1) interpret the design specification for a given device
(2) extract information from general design specifications
(3) read the physical, electrical, and mechanical specifications for a given system
(4) interpret and comprehend the operational specifications
(5) list all factors which should be considered for each component of general and operational specifications

INTRODUCTION

This chapter will introduce you to design specifications. Although we have titled this chapter design specifications, it does not mean you will actually design a system. The design process is a long and arduous job that may take years to complete. When you purchase a computer or programmable controller, the cost of developing the system is included. "What does this mean to me," you may be wondering. By examining a set of specifications, you can determine if certain design conventions have been followed. Competitive markets have made it imperative that equipment be reliable from design to final product, but that is no guarantee all companies will comply. By looking at the specifications for the equipment, you can determine this for yourself. Good design procedures mean reliability. Have you ever noticed that the longer a system is down the more managers you see? This is because downtime means not only lost production time but lost profit. By buying reliable equipment, you can reduce the mean time between failures (MTBF). Another way for avoiding downtime is to use a good preventative maintenance program.

We have divided design specifications into two sections: general design specifications and operational specifications. Many of the requirements involve engineering principles, while some are just good common sense. We will examine the general specifications first with sample specifications presented at the end of the chapter.

SECTION 1—GENERAL DESIGN SPECIFICATIONS

Physical

The physical part of the specifications refer to the actual shape or layout of the system. This usually means the PCB (printed-circuit board). There are four major considerations:

First, the physical size and shape of the PCB needs to conform to some bus standard. This means the frame in which the PCB is mounted will accept other printed circuit boards and have some capability to be interchangeable with a similar mounting system.

Connectors and connecting systems are the second area to examine. They should allow solid connections between the mounting frame and PCB. By standardizing the mounting system, we assured each card will be interchangeable.

The etchings on the board itself are the third area of concern. Line etchings are important when definite power requirements must be met and when overheating might occur. Improper arrangement can also induce unwanted electromagnetic interference or create stray capacitance.

The fourth concern is about the physical layout of the PCB. The PCB layout necessitates a great deal of engineering in order to assure that all functions can be mounted without interfering with nonassociated functions. Some of the design procedures are so complicated and time consuming that computerized drafting is used to locate and provide correct placement of components. Signal sources placed too close to potential receiving components will generate interference at the board level. By spacing them some distance apart most of this can be eliminated.

The height of the PCB can be very important also. By keeping the board and component height to a set value, mounting space in the unit can be allocated and adequate ventilation provided. A maximum height of 0.75 inches is not uncommon.

Properly securing components to the PCB is just good common sense. Loose components can cause system failures. Interchangeable chips need to have proper holders to allow for replacement; and good soldering techniques, such as wave soldering, may be far superior to hand soldering. Hand soldering can lead to what is called a *cold solder joint*. Although a component may look and measure properly with a continuity check, it may fail to operate once the PCB warms up or cools down.

Many times during the PCB development stages, problems are encountered and changes must be incorporated. If these are minor and involve merely placing a jumper wire between two points, a wire is hand soldered in place. This is called an *engineering change*. An engineering change is much more economical than a redesign process, especially if it means a total redesign of the entire PCB. Excessive engineering changes are also an indication of lack of proper design considerations. Final production products should have few changes, but a prototype system may have many changes.

Electrical

The electrical portion of the general design specifications can involve many possible areas. By incorporating the four areas we will examine many other techniques that must have been previously applied during the engineering process.

Buffers should be installed at any point of the system where potential timing variations and operating levels vary between two components. These serve both an organizational functional as well as a protective function.

Timing units need to operate at sufficient speed to allow data transfer and processing before an update is required. This may not be readily changeable to meet your needs and may become a factor in deciding whether or not to purchase a particular system.

Test functions are critical if safety is involved. When they function erroneously, computer systems need to disengage and fail to a safe condition that will prevent loss of life, injury or damage to equipment.

Another area to consider is Federal Communication Commission requirements to eliminate unwanted electromagnetic and radio interference. Proper shielding is the responsibility of the manufacturer. It is your responsibility not to remove or tamper with the shielding.

Mechanical

This area again falls into the category of common sense. Industrial environments are usually not very gentle when the care of electrical equipment is concerned. Dirt and vibration are a necessary part of many processes. Adequate measures must be taken to eliminate the effects of vibration and to keep parts from becoming disconnected. The use of snaps, hold-down clips or retainers is sufficient in most cases.

SECTION 2—OPERATIONAL SPECIFICATIONS

The second section discusses the operations specification. This specifies the operational requirement that systems must meet in order to survive in your environment. A set of operational specifications follows the design specifications. These specifications are not meant to fit every possible application but can be modified to meet your own specific needs. They represent the combined inputs from a variety of industrial situations including utilities companies, automotive manufacturers, appliance systems, food processing, electronic manufacturing, metal processing, plastic films, pharmaceuticals, robotics, communication equipment, and aerospace system parts.

Since many of these requirements have been discussed in some detail in Chapter 10, they will not be repeated.

SECTION 3—MOTOR CONTROL INSTALLATION

Installing the electrical components in a proper manner is the next logical step after a circuit has been designed. Major control circuits are usually manufactured with enclosures that are suitable for general purpose operation. If special conditions exist, optional units can be ordered to meet those needs.

In the previous sections of this chapter about design specifications, many important considerations for component selection were presented. Reviewing those will give you some insight into the special needs of electrical equipment such as programmable controllers and microprocessors. A general guideline for selecting and installing controls is first to choose the proper control size and then consider where it is to be installed. If the type of control or location creates a problem with expenses, it may be necessary to move it to another location—that might, for example, reduce these requirements and allow a less expensive unit to be used.

Equipment cost is many times the greatest obstacle in selecting and installing control equipment. Generally, the cheapest is not always the best, and there may be drawbacks for selecting equipment with this criteria. For example, a relay may be purchased for $50.00 when a superior unit is available for $70.00—on the surface it would appear that the $50.00 relay is most economical. However, considering the fact that the first relay may have a mean time between failures of 10,000 operations and the second unit 100,000, it should be obvious that the second unit may be a better buy. This would be an individual decision that must be made by the installer and the customer or by the supervisor in charge. Down time during replacement of the relay may also be another important factor. Hopefully you can see that there are many items that need to be examined for each individual case. The specifications available from the equipment suppliers can also aid in making decisions on what items to purchase or install.

Actual installation is another consideration. When replacing any component, it is generally best to replace it with an exact duplicate of the defective item. This assures that you stay within the design limits of the original and also maintains the appearance of the manufactured unit (same manufacturer, style, size, type, etc.). This may sometimes cause a problem for certain units that might require their replacement parts to be specially ordered. Depending on the nature of the problems created by the defective component, this can result in a long production delay or just a minor inconvenience. Assuming you have the appropriate replacement item on hand, then a logical procedure might be as follows:

1. Initiate any safety precautions necessary for the prevention of injury to personnel or equipment.
2. Examine the defective unit for possible causes of its failure. If the cause is not eliminated prior to replacement, you will probably experience another equipment failure.
3. After the cause is determined, repair or replace the broken item.
4. When using identical replacement parts, obtain the serial number, type, size, and manufacturer's name listed on the defective component. This provides for ease in ordering or purchasing a similar item. It may be necessary to consult a parts diagram if this information is missing or damaged.
5. After proper safety precautions have been taken, carefully remove the defective unit by following a step-by-step process that will allow all wiring and components to be reinstalled.

6. It may be helpful to tag or label wires on a relay or other component before they are removed. You might also consider making a diagram of that component and its associated wiring if none is available.
7. Replacing identical type items is usually just the reverse of removing the component. By saving the hardware (screws, clamps, and brackets) from the damaged units, the installation of the new unit will be made much easier.

There are other occasions when a new item must be installed for the first time. Again, as a general guideline, use parts from the same manufacturer if they are adequate and available. This helps to maintain the appearance of keeping equipment as a single unit instead of something that has been just thrown together. Appearances are important and should be considered along with all technical aspects of selecting the appropriate items.

Before you place a new item in a control box or cabinet, study the general layout of the items and follow the same scheme, if possible, and also allow for other additions or changes that might need to be incorporated at a later date. If no pre-existing unit is available, and you are to design your own equipment layout, begin by considering if this is a permanent or temporary change. Permanent changes should generally be placed within enclosures designed for the environment. This is the time to consult with your equipment supplier, equipment catalogues and the National Electrical Code (or your Local Electrical Code) for specific requirements on enclosures and wiring methods. This will assure that the equipment installation will be within the guidelines of what has been considered safe by industry and insurance company standards. If space allows, always consider placing many relays or starters within a single enclosure instead of having many individual enclosures. This not only allows for a neater appearance but provides additional space that may be used for future changes if needed. Enclosures add to the cost of any item. Installation time is also decreased, and to industry, time is many times the deciding factor on how changes must be made. Installing components in a cabinet should begin by placing them with a general layout that will leave adequate space to bend wiring, permit maintenance, enable cooling, and allow proper mounting for each item.

Wires should be labeled and bundled, when possible, to keep a neat appearance. Terminal strips may also be helpful in troubleshooting later and when many items are interconnected. A cabinet with a disconnecting means can also provide an added safety feature when you are working with it. Some units even employ door switches which act as safety interlocks to remove power when the door is opened, which can protect unauthorized personnel from injury should they gain access to the unit.

You may find it more convenient to mount devices by drilling and tapping the holes for the mounting screws. This makes it easier to remove and replace items, when necessary, and this also keeps the cabinet sealed when the screw is in place by not having oversize holes drilled for nut-and-bolt-type installations.

If you conform to the existing cabinet layout and try to install components in a neat and clean manner, you will not only have a more functional system, but one that will be beneficial later when it may become necessary to troubleshoot for a system malfunction. Neglecting to label wires, not keeping drawings of new installations or changes, and poor wiring techniques will only add to your problems, or those of your coworkers, when a system error occurs; and this is usually when you have the least amount of time to study how the equipment was wired earlier. No matter what technique you choose, do it in a safe manner that will protect you, your fellow workers, and the equipment. No job is so important that it cannot be done in a safe way that will protect everyone from harm!

SUMMARY

Design specifications are not meant solely for designers, and hopefully you have developed that idea from this chapter. By presenting some typical design considerations and applications, we hope you will be able to evaluate the limitations of both old and new equipment and specify the appropriate equipment for your own particular environment. Chapter 13 will discuss various power systems found in industrial plants.

GENERAL SPECIFICATIONS

I. POWER

Standard input voltage shall be 110 volts. Voltage fluctuations of ± 10% shall be tolerated. Transient voltage protection shall be provided at the power supply input. Unregulated voltages shall be supplied by the power supply and regulated voltage will be developed at the board level.

II. PHYSICAL

A. BOARD

The PCB shall comply with standard bus configurations.

B. CONNECTORS

Connectors compatible with the interface standard shall be securely mounted on the PCB or the cabinet.

C. LINE ETCHING

All etchings shall be of sufficient width to supply the maximum current required for the trace. Adequate spacing shall be maintained between power and ground traces. Long parallel traces are to be avoided when possible.

D. BOARD LAYOUT

1. Component layout shall be maximized to prevent capacitive coupling between components to the extent possible. Decoupling capacitors shall be placed adjacent to all EMI generating sources. EMI-sensitive components shall be located as far as possible from EMI sources.
2. Board and component height shall be a maximum of .75 inches to optimize space in the backplane and allow for air circulation.
3. All components shall be securely mounted to the board by wave soldering or inserted in appropriate component holders.
4. Engineering changes to the PCB shall be kept to a minimum.

III. ELECTRICAL

A. BUFFERS

Buffers shall be installed at all inputs, outputs, data and address busses to prevent overloading of the CPU.

B. TIMING

Timing circuits shall be of sufficient speed to be compatible with the sample rate of the process being monitored.

C. TEST FUNCTIONS

A periodic automatic self-test function shall be available on all PCB functions that are critical to safety. Upon failure of a self-test function, an error flag shall be initiated and a stop sequence begun as either an emergency stop, sequence stop, or coasting stop. Stop sequence shall be determined by the process under control.

D. FCC REQUIREMENTS

The PCB and the enclosed unit shall comply with current FCC regulations for EMI and RFI noise suppression.

IV. MECHANICAL

The board shall be structurally sound and capable of withstanding vibration and mechanical shock without dislodging the backplane connector. Screw snaps, hold-down clips or retainers shall be installed for this purpose.

OPERATIONAL SPECIFICATIONS

I. ENVIRONMENTAL CONSIDERATIONS

A. CORROSIVE

1. CHEMICAL GASES OR ACID VAPORS

Printed circuit boards exposed to chemical gases or vapors shall be enclosed in NEMA 12 metal enclosures with positive air pressure applied to the interior. Additional protection for the PCB shall be provided by coating the board with an inert film of polyurethane.

2. MOISTURE

Printed circuit boards exposed to moisture shall have a humidity control system on the enclosure, consisting of either ventilation, air conditioning, or a dehumidifier. Enclosures should generally be of the NEMA 12 type.

3. TEMPERATURE

a. Printed circuit boards shall be capable of operating in temperature ranges from 32 degrees F to 150 degrees F ambient.

b. Environments with high ambient temperatures shall have an additional 20% open space available for air circulation and ventilating fans installed to prolong PCB life. Air conditioning can be provided as an alternative.

B. NONCORROSIVE DUST OR DIRT

1. CONDUCTIVE

Printed circuit boards exposed to conductive dust or dirt shall have either a controlled environment within the enclosure or an air filtration system installed. Additional protection shall be provided by coating the PCB with an inert coating of polyurethane.

2. NONCONDUCTIVE

Nonconductive dust or dirt shall be treated as a conductive dust environment with the exception of the PCB coating.

C. EXPLOSIVE

Generally PCBs will not be placed within an explosive environment. If it is necessary, then suitable NEMA enclosures shall be used that satisfy the class and division of the environment.

D. VIBRATION AND SHOCK

Units subject to vibration or shock shall have sufficient shock mounting to prevent damage to the PCB, etching, or loosening of the cards from the edge connectors.

II. ELECTROMAGNETIC INTERFERENCE

A. INPLANT

1. POWER SYSTEM

Power supplied to microprocessor units shall be from isolation transformers with transient protection to eliminate voltage spikes. High noise environments may require additional shielding of the microprocessor unit within the enclosure. Enclosures of the NEMA 12 type are generally adequate.

2. MAGNETIC

Magnetic interference from welders and other sources shall be eliminated by any one or all of the following means: (1) provide separation from the source, (2) metal enclosure or shielding, (3) not operating magnetic devices near the microprocessor.

3. RADIO

Radio interference shall be eliminated the same as magnetic interference. An alternative to radio interference is using hardwired intercom sets on an additional pair of wires in the communication cable.

B. EXTERNAL EMI

External EMI problems shall be eliminated by first identifying the source, notifying the responsible agency, and using the methods for magnetic interference when possible.

C. WIRING SYSTEM

1. Interference due to the wiring system shall be eliminated by one or all of the following methods: (1) separating the signal and power cables, (2) placing the signal cable in steel or ferrous metal conduit, (3) shielding of signal and power cables when run in close proximity, (4) avoiding paralleling signal and power cables, (5) crossing cables at right angles when possible.

2. Two-wire twister pair is adequate for most data transfer over short distances. Shielding should be of the foil type with a drain wire.

3. Data transfer for long distances shall be coaxial cable or cable suitable for specific network systems.

D. GROUNDING METHOD

Ground terminations shall be connected only at one end of the signal cable to prevent circulating ground currents. This should be done at the microprocessor end of the cable to assure a common ground reference to all devices.

III. POWER SYSTEM

A. VOLTAGES

Power supply voltages shall be 110 volts AC.

B. FAILURES

1. BACKUP SYSTEMS

Battery backup systems of at least 6-month duration shall be available.

2. DEFAULT CONDITIONS

a. Upon power system failure, default conditions shall go to either an emergency stop, sequence stop, or a coasting stop condition. Stop sequence desired will vary with the process under microprocessor control.

b. Regardless of the stop method, a manual restart must be initiated before microprocessor control can resume.

IV. ENCLOSURES

A. TYPE

The standard enclosure shall be a NEMA 12 metal cabinet.

B. SIZE

1. The size of the enclosure shall be specified by the end user. Units designed to be free standing will generally be floor mounted. The microprocessor unit shall be capable of being mounted inside the enclosure with an additional space allowed for air circulation and future system additions.
2. Size and type are dependent on the standard practice of the user, the final size of the microprocessor, the controlled process, and the environment. Final decisions must be based on all of the above conditions.

C. LOCATION

Final location shall be determined by conditions similar to size and type. Controllers will generally be within the area or process under control.

V. EXISTING CONTROLS

All controls connected to the microprocessor shall be isolated by the use of optoisolator circuits with an isolation capability of 4KV or greater.

A. RELAYS

1. Relays installed as inputs and outputs on the microprocessor shall have transient and noise protection in the form of metal oxide varistors, snubbing circuits, or free-wheeling diodes.
2. Loading capability of each output or input must be specified to prevent exceeding microprocessor ratings.
3. Relay contacts shall be rated for the voltage and current supplied by the source. Replacement relays shall not be used that have been exposed to voltages higher than design voltage supplied by the source.

B. PHOTOCELLS

Photocells shall not initiate transients on startup or during relay or switch operation. Protective means used for relays shall be adequate.

C. LIMIT SWITCHES

Limit switches shall have the same consideration as relays for loading and transient protection.

D. STEPPING MOTORS

1. Stepping motors shall have the same consideration as relays for loading and transient protection.
2. Resolution and speed of the stepping motor shall be compatible with the process speed of the microprocessor and the information sample rate desired from the process under control.

E. PROXIMITY SWITCHES
Proximity switches shall be given the same consideration as photocells.

F. PROGRAMMABLE CONTROLLERS
Programmable controllers shall be considered the same as microprocessor units for input and output devices.

G. MICROPROCESSORS
Microprocessors used as final control devices shall satisfy input and output requirements of the host microprocessor.

H. MANUAL DEVICES
Manual devices shall be of the nonbounce type when used as an input to the microprocessor. In addition they shall also satisfy requirements for transient and noise protection the same as relays.

VI. FUNCTION OF CONTROLS
The following control functions need to be available either as a single unit or in conjunction with another unit:
Position Sensing
Motion Sensing
Speed Control
Temperature Control
Sequence Control
Pressure Sensing
Level Sensing
Safety

VII. SAFETY OR DEFAULT CONDITIONS
Meeting the OSHA standards is the responsibility of the end user. However, controlled-stop sequences should be incorporated when possible, and any or all additional steps required by the user shall be specified as the user's responsibility.

VIII. MAINTENANCE
All in-plant maintenance shall be possible to the level of PCB replacement. Any additional maintenance beyond this level shall be done at OEM level.

IX. SOFTWARE REQUIREMENTS
A. INPUT DEVICE
1. TAPE
Tape drives generally are adequate for most applications that involve programmable controllers.
2. DISK
Disk drives are acceptable only in locations that are not considered to be harsh environments.

B. LANGUAGE

Language used in programming shall be user defined. Possible languages include Fortran, Pascal, and Basic.

C. FORMAT

1. Format for programmable controllers shall be a ladder diagram.
2. To the extent possible programs shall be menu driven at the development level.

D. DIAGNOSTICS

1. Self-test diagnostics shall be incorporated during start up and as a periodic function of the program. Specifications shall comply with test function requirements.
2. Identification of component or rung-level failures shall be available as well as excessive time-in-loop failures.

E. PROGRAMMER

Programmers shall be determined by end user. Security for the program shall be provided by key lock or password.

F. TRAINING

Training shall be supplied by the OEM source to the end user on an in-house basis. Documentation for training shall be adequate to allow further in-plant training utilizing personnel attending the initial training session.

EXERCISES

On a separate sheet of paper, complete a response for each question, statement, or problem listed below.

12.1. The result of a good design procedure is ______.

12.2. Lost production is not the only consequence of long downtime, but also ______.

12.3. An alternative way of preventing downtime is to use ______.

12.4. What are the two main sections of design specifications?

12.5. List the three major classifications of the general design specification.

12.6. What does a physical specification convey?

12.7. What does the electrical design specification involve?

12.8. List the four major components of electrical specification that must be examined during the design process.

12.9. What factors do we consider in the mechanical design of a device?

12.10. What are operations specifications?

12.11. List the six major points you would consider as software requirements of the operational specifications.

13

Power Systems

OBJECTIVES

Upon successful completion of this chapter, you should be able to

(1) describe the different power generating processes
(2) describe what a generator is and how it operates
(3) explain power distribution methods
(4) describe how a single-phase transformer works
(5) explain the operation of a 3 phase transformer when it is connected in a star-star configuration, delta-delta configuration, delta-star configuration, or star-delta configuration
(6) describe the use of capacitors in power-factor improvements and their significance for the increase of a power system's efficiency

INTRODUCTION

To work successfully with any control system, you need to understand your power source. Not everyone will work at a power generating station, but many high voltage sources are present at the plant level. This chapter will discuss power systems from the power plant level and then trace it to the control level.

UTILITY POWER

Power plant structures are developed by a combination of geological and economical factors. Some of the various types of power plants or generating stations are fossil fuel, hydroelectric, thermoelectric, and nuclear. These classifications indicate the source of energy used to power the electrical generator.

Fossil fuel plants use coal or some other combustible fuel to generate steam which in turn is used to turn the generator. This type of plant can be found in almost any part of the country. The presence of large coal reserves makes these plants very economical, but they do have some disadvantages. Some of the byproducts of the combustion process are hydrogen sulfide and fly ash. Hydrogen sulfide has been linked to acid rain and fly ash to air pollution. To eliminate these byproducts, additional cleaning units called *scrubbers* are added. They clean the air before emitting the exhaust gasses into the environment. The added expense of this equipment is high and reduces the efficiency of the entire process. This, of course, translates to increased operating cost and higher utility bills.

Hydroelectric plants are found in areas with a natural flow of water either as a river or a man-made lake. These are probably the cleanest sources of power available but are dependent on an adequate water source. When new plants are built, environmental factors must be considered, as well as a need to obtain large tracts of land.

Thermoelectric or geothermal plants are not as common, since they must be located close to geological faults which produce hot gases near the earth's surface. The hot gases in turn are used to develop steam to turn generators.

Nuclear plants are the most independent plants as far as location requirements but remain the most controversial because of the nuclear waste products developed. Nuclear reactions generate large amounts of heat that are used to make steam that powers the generators.

Although the source of the energy may vary, it is at this point that all electrical generators begin to have a common reference, the generator. The generator has two major components, the turbine and a rotating electric field. The turbine is forced to rotate by a mechanical force applied by either high pressure steam or water pressure. The turbine rotates a common shaft with the electric field, and this is where electrical energy begins its journey from the power plant to your control system. Figure 13-1 summarizes the first portion of the generation cycle at the power plant level.

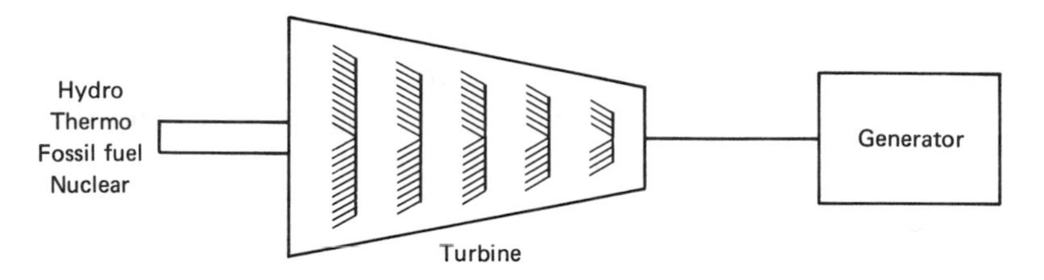

Figure 13-1 Electrical power sources.

GENERATOR

The generator consists of a variable magnetic field between two poles, a loop of wire, and a set of slip rings. Figure 13-2 shows a diagram of a simple generator. Actual power plant generators do not physically look like our unit, but they operate on the exact same principle that will be discussed. Let's begin by examining the necessary actions for electrical generation and then the purpose of each component.

Three essential elements must be present to generate an electrical current or voltage: a magnetic field, a conductor that completes a circuit, and a relative motion between the two. These are all present in our simple generator diagram.

The two magnetic poles supply the magnetic field. By changing the position of the rheostat, the magnitude of the field can be varied. The loop of wire has a current induced into it by rotating through the magnetic field. The slip rings allow

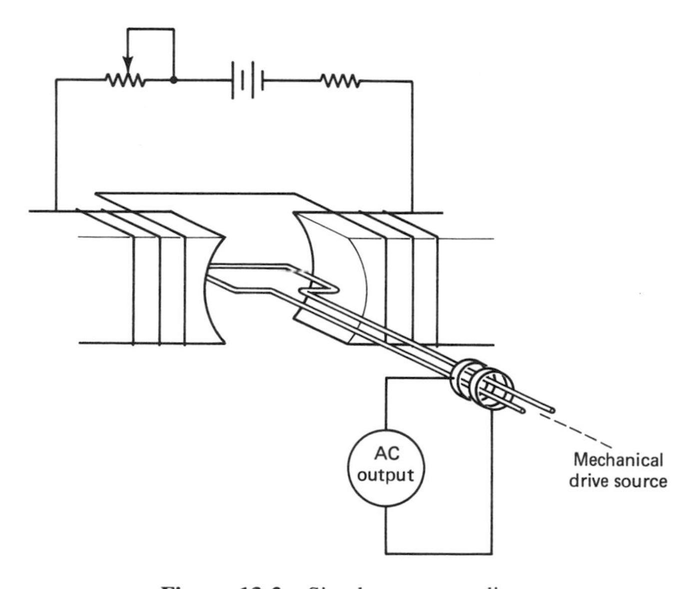

Figure 13-2 Simple generator diagram.

a movable connection to complete the circuit to a load device. The mechanical rotation of the loop is all that is required to generate a current. It is the physical arrangement of this loop in the field that is the control for generating the alternating current that is our standard source of power. Figure 13-3a examines the relationship of this rotating loop in a generator. A cross section of our generator is represented in Figure 13-3b.

The solid arrow lines are the lines of force developed by the magnetic field, and the dashed circle is the path of the wire loop in the magnetic field. In this example the mechanical and electrical zero reference points exist in the same point as shown. To measure the amount of current generated, the angle created by the loop as it crosses the line of force is used. Figure 13-3b shows this angle as vector-drawn tangent to the rotation path at the point where the loop and a line of force coincide.

The angle at which the wire crosses a line of force determines the strength of the downward vector at that point in time. An analogy might be the amount of downward force developed by dropping a rock off a pier. If you drop it straight down and perpendicular to the surface of the water, the maximum amount of force is transferred. If you try to skip the rock across the surface at some small angle, the amount of force developed is greatly reduced. This is illustrated in Figure 13-4.

We are fortunate that a simple mathematical formula can be used to represent the force or magnitude of electrical current at a given point of the arc.

This formula is $I = I_{max} \times \text{Sin } \emptyset$.

Figure 13-5 shows the relationship of mechanical position of the loop and the electrical current developed. If this current is connected to a load resistance, a similar relationship can be used for voltage. This same mathematical relationship is what gives the alternating current the name sine wave.

A typical generator is more efficient when it is used to make three separate

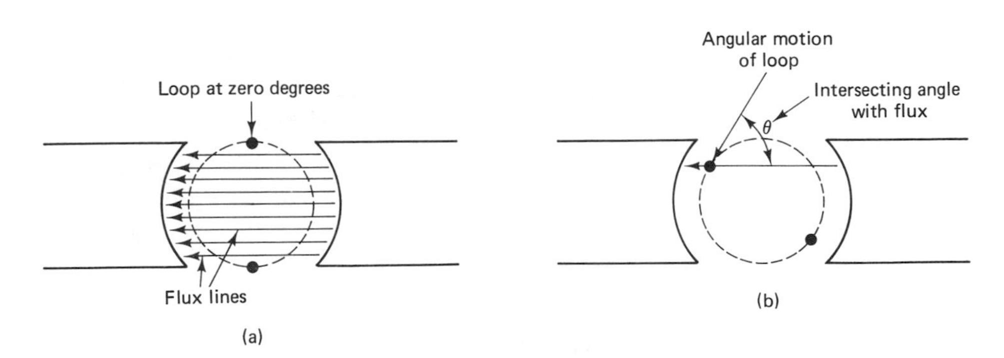

Figure 13-3 Rotating loop.

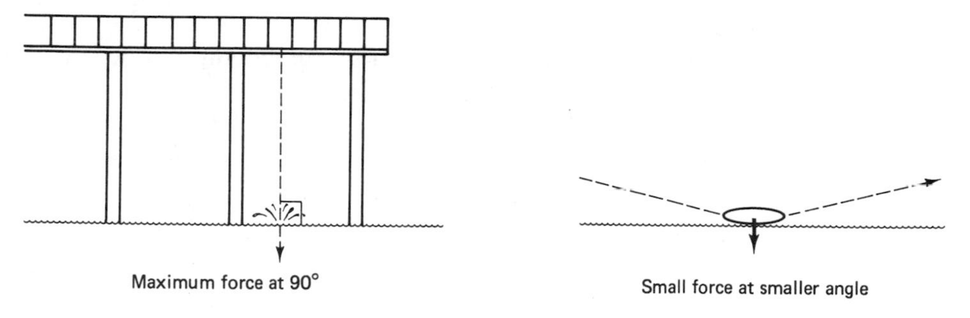

Figure 13-4 Angular force analogy.

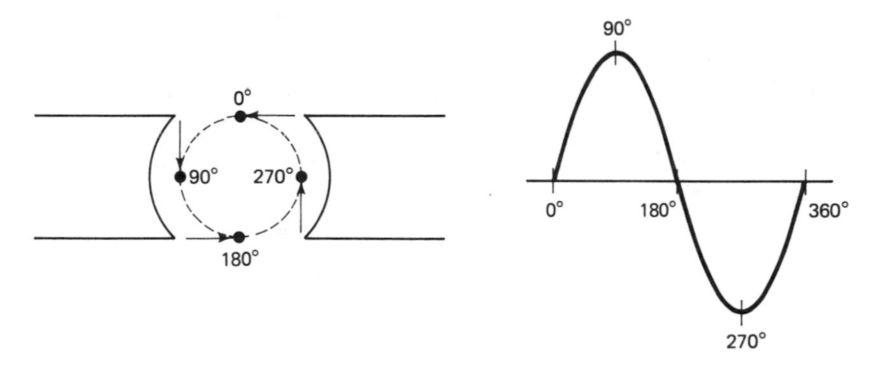

Figure 13-5 Sine-wave generation.

Figure 13-6 Three-phase generation.

outputs called *three-phase power*. The three phases are usually labeled phase A, B, and C and appear as shown in Figure 13-6.

DISTRIBUTION SYSTEMS

Once the power has been developed at a generating plant, some method needs to be used to transport this power to the user. These networks are called *distributions systems* or grids. The current from the generator is sent to a transformer which increases the voltage and then through an array of switches to transmission lines that carry power to intermediate substations that either increase or decrease the voltage, depending on whether or not the final user point is close. The main components of a distribution system are the transmission lines and the transformers. Switches and capacitors are also important, but we will limit our discussion to the transformers that are used.

Transformers

A brief explanation of transformer operation might be helpful at this time. Figure 13-7 shows a simplified diagram of a step-up transformer. The term step-up means the voltage on the primary side is stepped up on the secondary side. Alternating current on the primary side induces lines of force into the iron core.

The secondary of the transformer is wrapped around the same core and has many more turns of wire. The magnetic flux that flows in the iron core induces a current into each loop and, due to the large number of loops, the voltage is greater on the secondary side. It is the relative motion of the magnetic field and the conductor, similar to the generator, that makes the transformer work.

By reducing the number of turns of wire on the secondary in relation to the primary, the voltage can be reduced on the secondary side. In mathematical form this is expressed as $Vp/Vs = Np/Ns$. The ratio of the voltages of the primary and secondary are proportional to the turns ratio of the primary and secondary.

Figure 13-8 illustrates some typical voltages that are developed during the transfer of energy from the power plant to a residential area and a factory. High voltages (138,000 V) are used on the transmission lines because there are less current requirements, and this means smaller wires can be used. This is important because the larger wires are more expensive to purchase and require both larger and more supporting structures. The second substation in the diagram may serve to increase voltages to compensate for voltage losses due to the resistance of the transmission lines. This is usually required at each substation since the distances between substations cause further voltage drops.

Substation three has an intentional voltage reduction to bring voltages within the range of equipment that must be run at the plant level. Additional transformers at the plant level are used to adjust voltages to user requirements. Substation four lowers the voltages one more time to make them compatible with residential

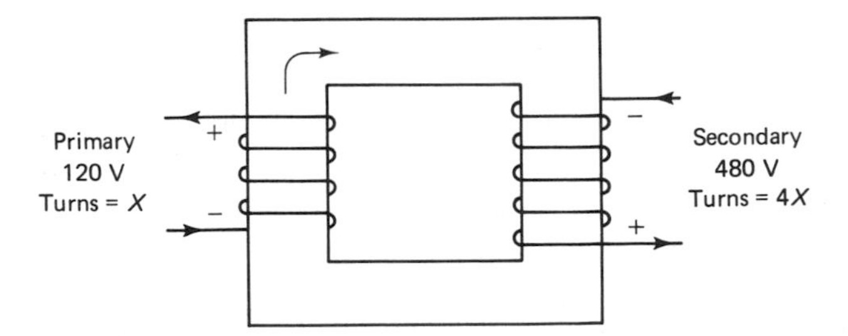

Figure 13-7 Simplified transformer.

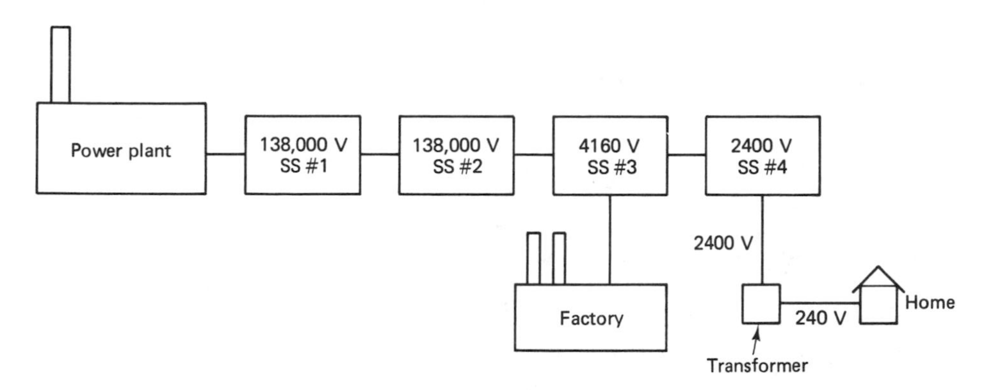

Figure 13-8 Power distribution system.

needs. The 2400-V level is high enough to prevent excessive losses but makes the use of a final transformer very convenient to drop voltages to the 240/120-V level that is found in many homes.

In-Plant Systems

Power systems are sometimes designated as single- or three-phase systems, but power companies generate only three-phase power. The single-phase power is actually one phase of three-phase power. It does simplify things somewhat when we look only at the power we use, especially at the residential level. With this in mind let's look at single-phase power as it is found in a residential environment.

Single-Phase Power

Substation four in Figure 13-8 has a secondary voltage of 2400 V. By using a step-down transformer this voltage may be reduced to 240 V.

Figure 13-9 illustrates this and shows several important ratings of transformers. The notation, 100 KVA, is a power-handling capacity of the unit and varies with power requirements. One unique feature of this rating is important. The rating is the product of the voltage and current of the primary or the secondary of the transformer. The rating of the primary must equal the secondary. This is very helpful when either voltage or currents are known for a given transformer. Consider a 100 KVA rating with 2400 V on the primary. This means the current present is 41.67 A. With this low magnitude, small wires can be used to supply this current to the primary. The secondary also has the same 100 KVA rating. Using this as a basis for calculations, the secondary current will be 416.47 A and the voltage is 240 V. This is more than adequate for most new homes, since typical power requirements are in the range of 200 A.

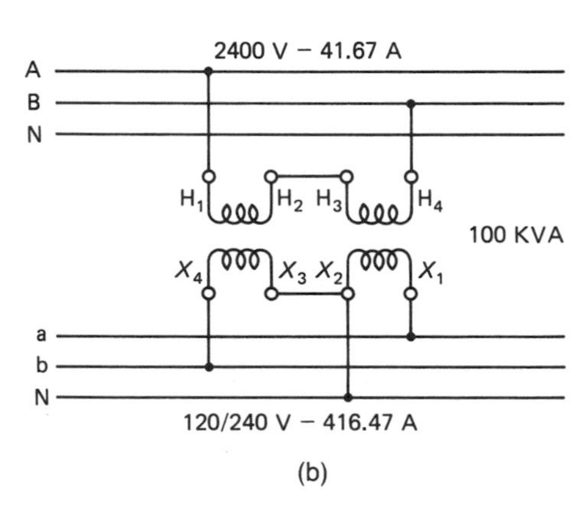

(a) (b)

Figure 13-9 (a) Three-phase transformer bank (courtesy of Square D Company). (b) Single-phase transformer connection (basic unit for three-phase transformers.

You should also notice there is a third wire call a *neutral*, which is normally grounded on one end and attached to the midpoint of the transformer, which helps develop a 120-V potential from each end of the transformer secondary to the ground point. This arrangement is very useful because large appliances that require 240 V can be connected as well as 120-V appliances. The loading requirement of a device is usually a measure of how much current it requires to operate. Although there is a mathematical relationship with Ohm's Law to calculate the voltage and current, the current is the basis for sizing the power requirements of a residence.

Three-Phase Systems

The three-phase power system is economical for both the power company and large industrial firms. Substation three in Figure 13-8 shows a typical voltage that might be used to power a factory's own substation. The higher voltages present offer substantial savings in energy usage when large power requirements exist. Not all equipment operates at these high voltages, and transformers are again used to reduce the voltages down to more practical levels. Instead of using a single transformer, three transformers called *banks* are often used to reduce all three phases at the same time. The various connections that are possible offer many different voltage and current combinations.

Star-Star (Wye-Wye) Configuration

A very common transformer hookup is the star-star arrangement shown in Figure 13-10. The wiring diagram for the transformer bank is illustrated in the top portion of Figure 13-10. Most transformers used have dual windings that are connected in series or parallel to adjust for different current or voltage levels. The primary side is labeled using the letter H for each terminal and the secondary uses the letter X. Using this marking system, H1 and H2 would represent one winding of the primary, and H3 and H4 the other primary winding. A corresponding system is used for the secondary. The turns ratio is used to determine the change in winding voltage from primary to secondary.

Voltage measurements are done in two ways, phase-to-phase, and phase-to-neutral. Although the neutral is shown grounded in this diagram, this is not always a requirement. Many times, the voltages possible on one side of the

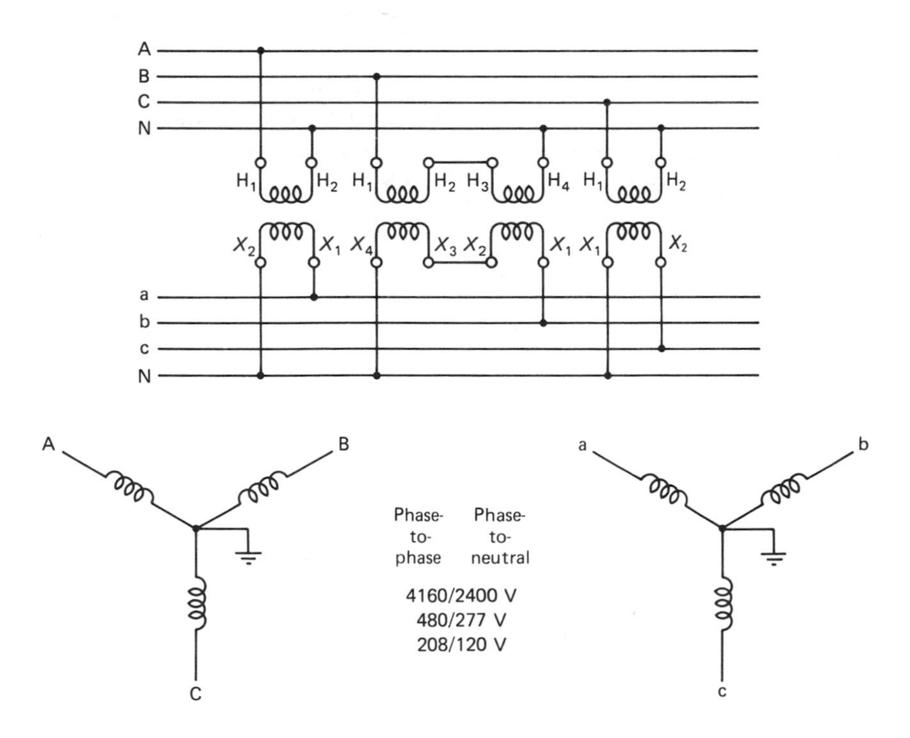

Figure 13-10 Star (Wye) configuration.

transformer are expressed as two voltages. For example, 4160/2400 V, indicates a phase-to-phase voltage of 4160 V and a winding voltage of 2400 V will be found in this system. A balanced star system, in which all three phases have the same load, has a mathematical relationship expressed as $V_{line} = 1.732 \times V_{winding}$. Other voltage arrangements used are given in the lower portion of Figure 13-10. Try using the formula given to verify the voltage relationships. You will find they are not exact but are close enough to allow reasonable voltage and current calculations. Since a small percentage of the voltage of each phase is present on each winding, the star connection is suitable for power transmission, gives three-phase voltages, and can also be used for lighting circuits as well. Keeping the system balanced can be a problem, and this is its major disadvantage. It is also subject to producing unwanted harmonics of the generating frequency which can interfere with audio equipment.

Delta-Delta Configuration

When a large three-phase load requirement exists with a small lighting load, the delta-delta system is often used. Figure 13-11 illustrates this hookup. The voltages from phases B to C provide 240-V power with a centertap for 120 V.

This arrangement is best for a small business with large three-phase motor loads that require additional power for office lighting and equipment. The disadvantages are that the lighting load will unbalance the system and every third circuit has a high voltage to ground. This voltage is 208 V. Anyone not familiar could inadvertently connect a 120-V load to this part of the circuit and destroy the equipment. For this reason the high leg is usually marked with a red or orange label to identify it from the 120-V connections.

Delta-Star Configuration

There are times when a compromise must be made between power and lighting circuits. The delta-star connection shown in Figure 13-12 is a good example. The increase in voltage available from the star connection, as well as the ease in balancing the load, makes it a popular configuration. The reverse version, the star-delta connection, lends itself to power connections but sacrifices the lighting load to achieve this benefit.

Open-Delta Connection

One of the additional features available with the star-delta connection is that in the event one transformer becomes defective, a hookup that electrically appears to be a complete three-phase connection can be used to power equipment during a

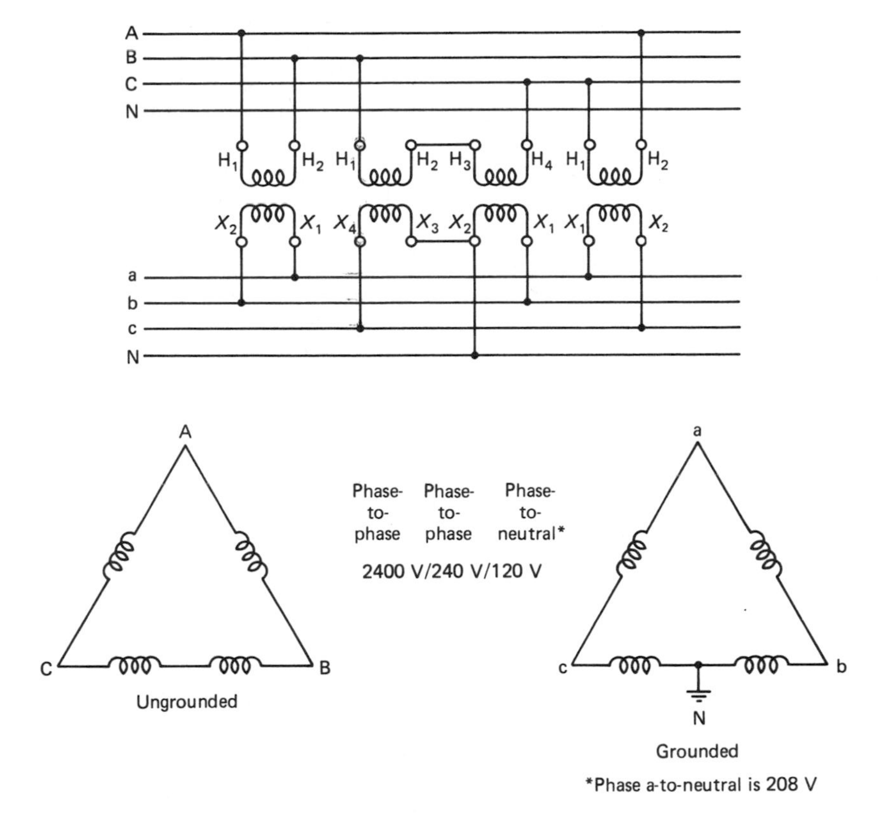

Figure 13-11 Delta Configuration.

power outage. Power loads must be reduced to compensate for the reduced capacity of the two transformers, but this is a minor inconvenience compared with a total power outage for an extended period of time. Figure 13-13 shows this connection.

Voltage-Current Relations

To calculate the voltages and currents in the transformer, a few basic formulas are used. Figure 13-14 is a star connection with its voltage and current relationships. Voltages are increased with the star connection while currents remain a reference point.

Figure 13-15 shows the delta connection. Voltages are the references in this circuit, and currents are increased. Power used in both of these circuits is the product of the voltage and current times the power factor of the circuit. A factor of 1.732 must be multiplied to allow for phase relationships in three-phase circuits.

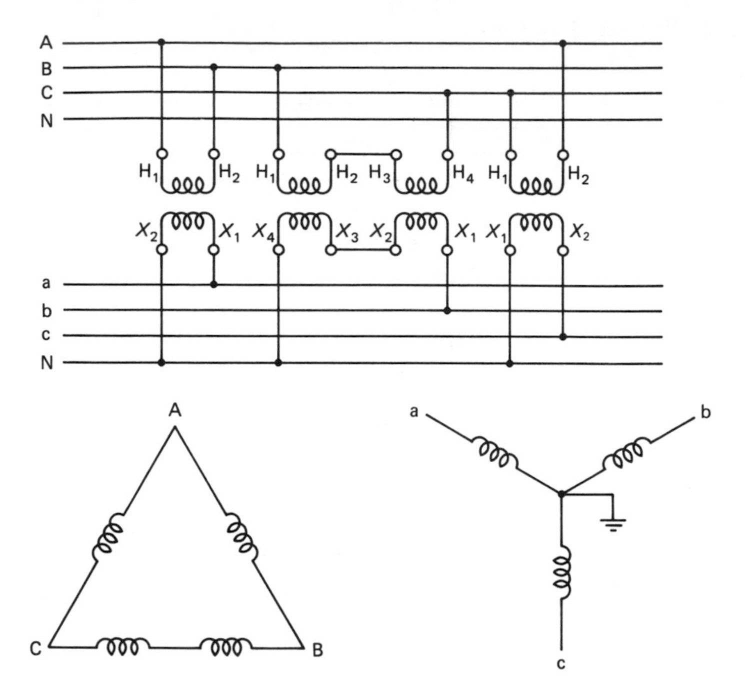

Figure 13-12 Delta-star configuration.

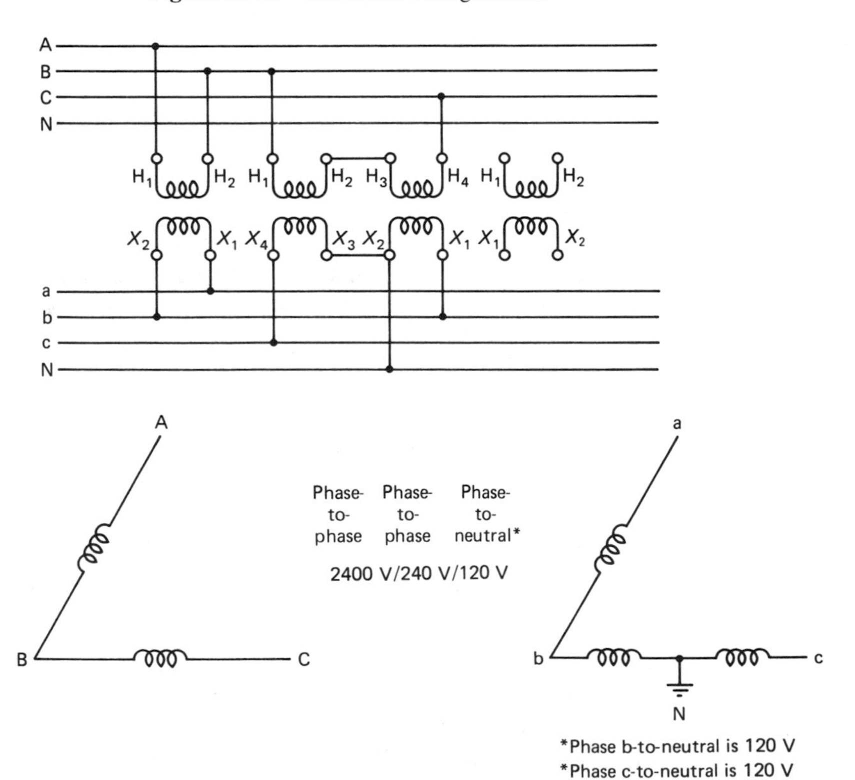

Figure 13-13 Open delta connection.

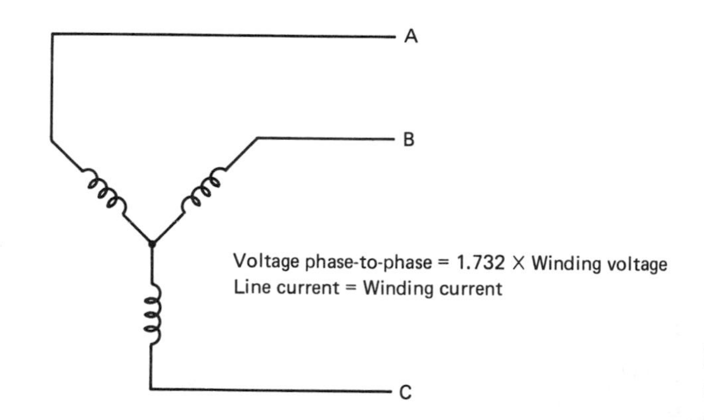

Figure 13-14 Star connection relationships.

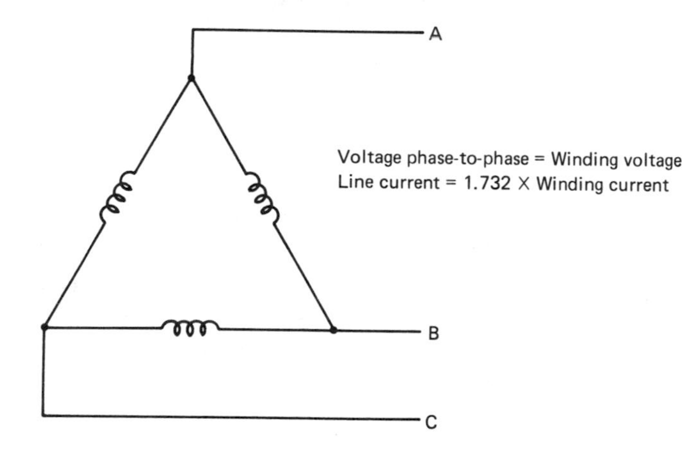

Figure 13-15 Delta connection relationships.

Power factor is a measure of the amount of reactance created by inductive and capacitive loads. As the net reactance is increased, the efficiency of the power system is decreased. Since motors have inductive reactance, they have a major effect on the power factor of a circuit. Capacitors are sometimes added to heavy motor load circuits to counteract this inductive reactance with capacitive reactance. Economically, the ideal value of having capacitive and inductive reactances equal in magnitude may never be reached, but it can produce a great savings to both the plant and the power company. Your utility company can aid you in determining the amount of correction required to eliminate this problem. They may also impose a penalty on your utility bill if you do not correct it.

The ratio of the reactance to the resistance in your circuit is given a numerical value that corresponds to the cosine of the angle formed. For this reason power factor is often represented as the Cos Θ in power formulas. Figure 13-16 is a diagram of the relationship.

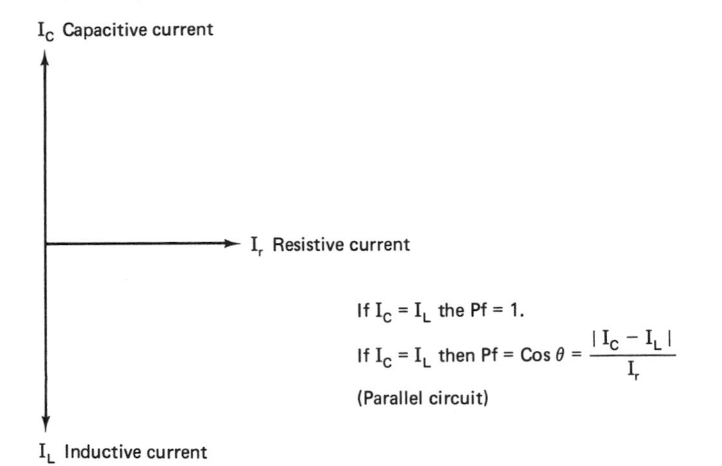

Figure 13-16 Power factor.

SUMMARY

This chapter has provided an overview description about power systems from the power plant level to the industrial and residential level. The voltages obtained can be reduced to 120 V, which are typical control voltages. Further reductions can be obtained by using additional transformers. Knowledge of the type of transformer arrangement and voltages present will allow you to add new equipment to existing circuits and obtain the correct voltages when needed.

Voltages up to 480 V can usually be measured safely with a good-quality tester. Never try to measure voltages above this level unless a high-voltage tester is used. These high voltages are very dangerous and can lead to permanent injury and even death. Never take a short cut around a safety cutout or procedure. Circumventing these procedures and using improper equipment can make you part of a short circuit. The values of short circuit currents will be discussed in the next chapter.

EXERCISES

On a separate sheet of paper, complete a response for each question, statement, or problem listed below.

13.1. There are four main classifications of power plants or generating stations. List them.

13.2. The raw material used in fossil fuel power plants is ______.

13.3. What makes a fossil fuel plant more economical and most common?

13.4. List two disadvantages of the fossil fuel power plants.

13.5. Which of the four power plant classifications requires geological faults for its operation?

13.6. Why is a nuclear plant the most controversial of all power plants presently available?

13.7. What is the common denominator among all power plants?

13.8. Name two major components of a generator.

13.9. Where in the power plant does the electric energy begin its journey to your control laboratory?

13.10. What is a generator comprised of?

13.11. Name the essential ingredients necessary for generating a voltage or a current.

13.12. The formula $I = I_{max} \times \text{Sin } \varnothing$ is used to determine the magnitude of electrical current at a given angle, Ø. At what angle will the magnitude be a maximum?

13.13. If $I_{max} = 10$ A and $\varnothing = 75°$, what is the magnitude of the current at this angle?

13.14. How does the efficiency of three-phase power compare to that of single phase when a typical generator is used for both?

13.15. What is the function of the intermediate substations?

13.16. What are the main components of a distribution system?

13.17. What does a step-up transformer do?

13.18. What is the basis for sizing the power requirements of a residence?

13.19. Write the expression for the line voltage of a star-star configuration.

13.20. What is a major disadvantage of a star-star configuration?

13.21. Which configuration would be best for small businesses with large three-phase motor loads?

13.22. What is the major advantage of the delta-star configuration?

13.23. What is a *power factor*?

14

Short-Circuit Protection

OBJECTIVES

Upon successful completion of this chapter, you should be able to

(1) describe the operation of short-circuit protection devices
(2) discuss what fuses are and how they are used
(3) explain fault current and fault current magnitude
(4) discuss fault isolation and analysis
(5) describe per unit systems
(6) explain impedance as it relates to generators, transformers or power lines
(7) apply short-circuit protection in a design situation

INTRODUCTION

One of the most common control devices that is changed or replaced in all electrical systems is the short-circuit protection element known as a fuse. When a current larger than the design current of the fuse passes through it, it opens up and stops the flow of current. Fuses are therefore short-circuit interrupting devices. Too many people have the notion that when a fuse continuously blows every time it is replaced the best thing to do is put a bigger fuse in the circuit. The only time this is true is when it has been underrated during installation. If the correct value was selected, then the fuse blowing is an indication of a problem that still remains in the circuit. Unfortunately, many people are satisfied with simply replacing the existing fuse without any regard to why a specific value was selected. As we mentioned in the previous chapter, short-circuit currents can reach high values. A term used to describe such a short-circuit condition in a circuit is *fault*. Some basic knowledge of fault currents is necessary to select fuses properly. The purpose of this chapter is to make you aware of the magnitudes that fault currents can reach and how to select a fuse based on data prepared by fuse manufacturers.

PURPOSE OF FUSES

Short-circuit interrupting devices are used to protect the generating company's equipment. There is no argument that these same devices protect the loads to which they are connected, but an ultimate purpose of these devices is to isolate the short or fault from the power supplier's system. It is theoretically possible for a fault in a home to shut down a power plant. Although this is an extreme example, the point is that for a power system to provide continuous and reliable service, some means must be provided to isolate the millions of possible fault conditions that are possible at any one time. Circuit breakers and fuses provide this function.

Figure 14-1 shows some of the symbols used to represent fuses and circuit breakers. Fuses are normally replaced when they blow, and the old one is discarded. Some types are available in which the link can be replaced. The link is the metal connection that completes the circuit between the two end caps. It is made to react (melt quickly) when a certain maximum level of current at a given voltage is reached. Circuit breakers are mechanical devices that react to excessive current and then open to break the circuit path. Low-current units use a bimetallic element that bends as more current flows through it. At its rated current value, this element releases a switch and opens the circuit breaker current path.

Magnetic units operate a small relay to perform the same function. They may be reset and used again instead of throwing them away or replacing a fuse link. This makes them more convenient than fuses, but not necessarily more economical. A properly selected fuse will not blow unless a fault condition exists. Unless the circuit develops many faults, using a fuse is much cheaper. (In this case, it would be necessary to maintain an adequate inventory of fuses. This is when the higher price paid for the circuit breaker becomes more feasible as well

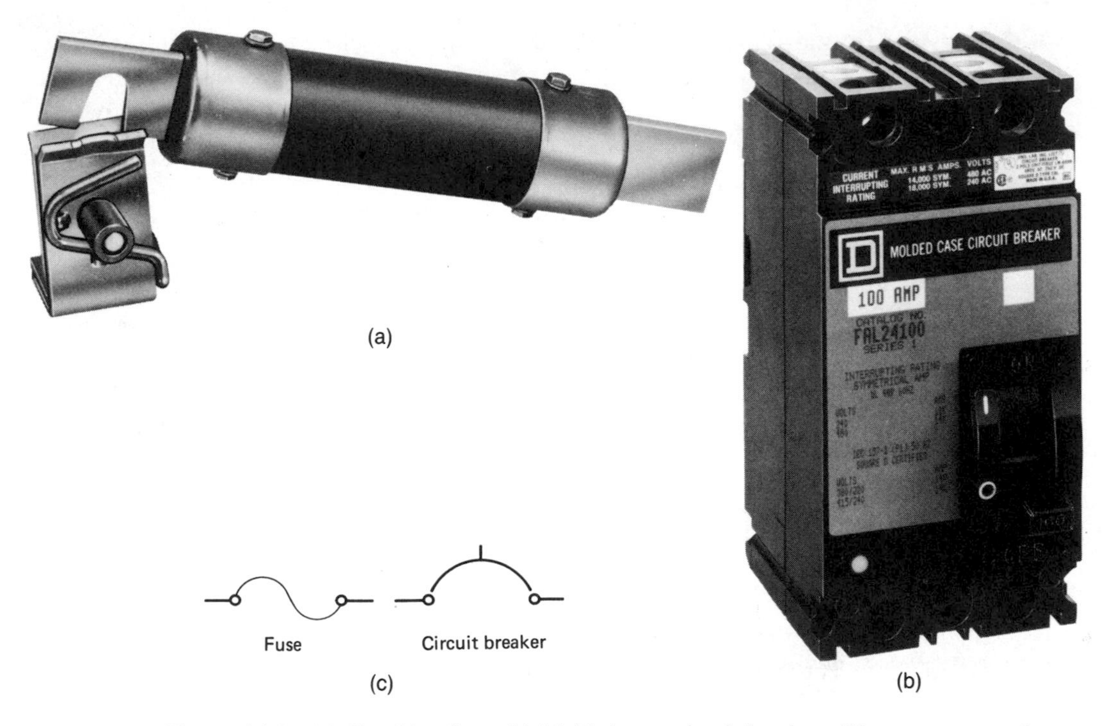

Figure 14-1 (a) Cartridge fuse. (b) Molded case circuit breaker. (Photos courtesy of Square D Company.) (c) Symbols for short-circuit protection device.

as convenient.) Other than the features mentioned here, the function of the fuse and circuit breaker is the same. The discussion of fuses will therefore apply equally well to circuit breakers.

FAULT CURRENT DAMAGE

When a fault develops, a large amount of current flows. This large amount of current can cause several events to take place. An arc occurs as the element in a fuse opens. This appears as a bright flash of light on fuses that have a window, but is concealed in the enclosed unit.

1. The heat that is produced can melt any surrounding metal that comes in contact with the arc. The larger the magnitude of the current, the greater the area of potential damage.
2. Magnetic devices such as transformers can also experience structural stress due to the physical movement of their internal parts.

3. Cables on transmission lines, in conduits, and cable trays will push away from each other because of the interaction of the large magnetic fields of the wires that are in close proximity to other conductors. Combine this effect with bad weather conditions and aging conductors, and the stress may cause the conductors to break.
4. One cycle of fault current may cause a conductor to break if the fault current value is of sufficient value to melt the conductor.
5. Insulation will break down because of the excessive temperatures or be weakened by stresses that are created by the fault. The insulation on conductors may also melt.
6. Current transformers will produce excessive amounts of current and can damage other equipment.
7. A fault will clear itself by literally destroying the material which provides the path for current to flow. This is undesirable—one of the purposes of the fuse is to allow for a controlled clearing of the fault.

Any one or all of these events may take place. The total effect of all the events is a function of their magnitude and duration. The longer the duration, the greater the influence.

Definition of Fault

A short or fault is any abnormal condition which provides an unintentional path for current to flow. The severity of damage caused by a fault is a function of the duration of the fault and the strength of the energy source. A properly selected short-circuit interrupting device will limit the duration of the fault and dissipate the energy that is generated during the fault.

Fault Current Magnitude

To develop the concept of fault current magnitude, normal circuit parameters should first be understood. The underlying principle is Ohm's law. The circuit in Figure 14-2 shows a normal circuit with a 5-A load. The basic components include a generator, a series impedance and the load impedance. The series impedance represents the sum of the generator reactance and its internal resistance and the line reactance and resistance. These factors are normally fixed once a system is installed. The current drawn by the load impedance will remain constant as long as the impedance does not change. Figure 14-3 illustrates the circuit changes when a fault occurs across the load. Two important changes have occurred.

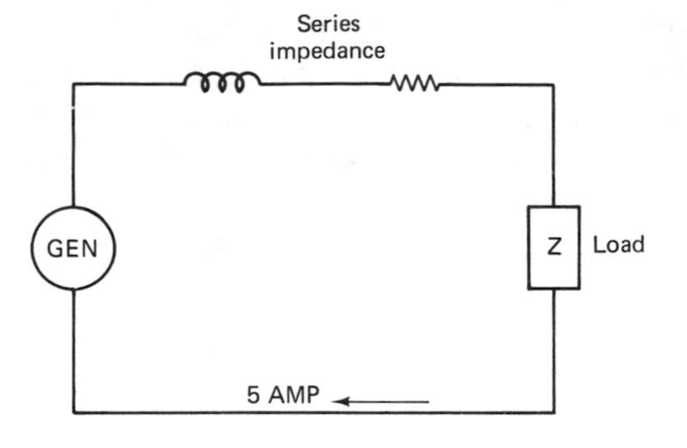

Figure 14-2 Typical circuit without a short-circuit.

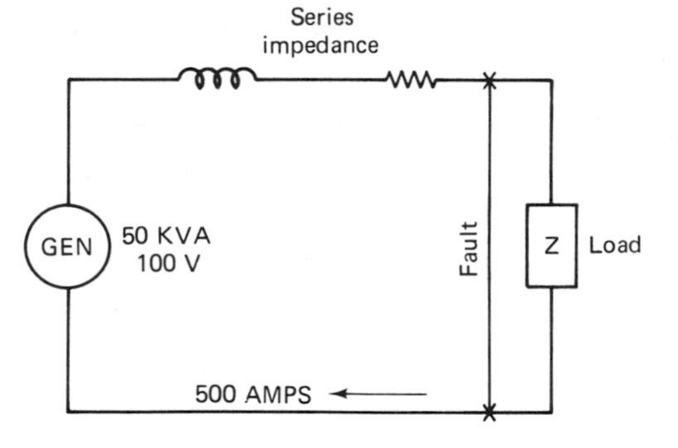

Figure 14-3 Circuit with a fault.

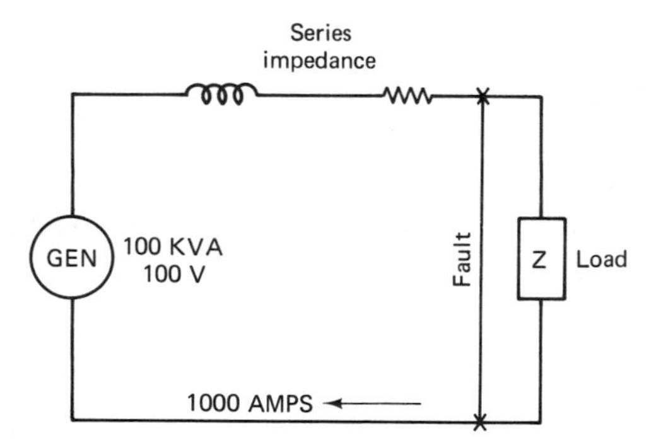

Figure 14-4 Fault current at 1000 A.

1. The load impedance is reduced to zero ohms, and the generator must now deliver its maximum rated current. Since the typical series impedance is a low value, the generator power determines fault current values. If the generator is rated at 50 kVA then at 100 V, the fault current is 500 A.
2. Increasing the generator rating to 100 kVA, as illustrated in Figure 14-4, increases the fault current to 1,000 A.

The potential magnitude of a fault current depends on the generator. It is not practical to calculate all of the reactances and impedances of a system clear to the power plant. A more realistic method is to consider the last transformer that is connected to your system to be a generator. This will be covered later in the chapter.

GENERATOR IMPEDANCE

One of the factors that determines series impedance is generator impedance. The generator impedance is composed of the generator's resistance and reactance. These values are based on the generator design and are constants. Each generator has its own characteristics. An in-depth study of fault currents requires knowledge of these constants, but for the purpose of this study it should be sufficient to mention that the subtransient reactance (X''), the transient reactance (X'), and reactance Xd are three of the important constants that are included in generator design. The subtransient reactance and transient reactance are usually used to calculate fault currents.

Transformer Impedance

A second component of the series impedance is transformer impedance. This impedance can greatly affect fault currents. To explain transformer impedance, the following analysis will be useful.

1. Considering all other values constant, if the transformer impedance is 5% with a 50 kVA rating and a secondary voltage of 240 V, the secondary current is 208.33 A. Since the transformer impedance rating is an indication of what percentage of the primary voltage is required to produce full-rated secondary current, this means 208.33 A is only 5% of the maximum possible short-circuit current. The maximum short-circuit current possible, as illustrated in Figure 14-5, is then 4166.6 A.
2. If the same transformer had a 2% impedance rating, as shown in Figure 14-6, then 208.33 amperes is only 2% of the maximum short-circuit current. A maximum current of 10,416.5 A is possible.

Transformer impedance is therefore an important consideration for initial short-circuit calculations. The transformer will be the generator for most applications.

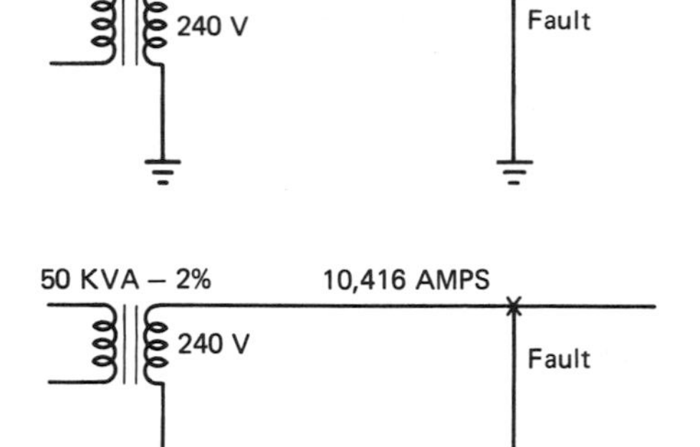

Figure 14-5 Fault currents at 5% impedance.

Figure 14-6 Fault currents at 2% impedance.

Line Impedance

Another component of the series impedance is the line impedance. This is the resistive value of the cables and the reactances associated with the interaction of magnetic fields from parallel conductors in the transmission system and raceways. Once the cables have been installed, these will remain a constant value.

PER UNIT SYSTEM

Power system values are calculated using *per unit* values. A per unit value is calculated using some circuit value as a reference base for all calculations. A 100-KVA rating referenced to a 1000-KVA rating would be 100KVA/1000KVA or 0.1 per unit (p.u.). Once the circuit calculations have been completed, they can be converted to their standard value by multiplying the per unit value by the base value used in the calculations.

FAULT ANALYSIS

There are three forms of current used in fault analysis: normal symmetrical current, symmetrical fault current, and asymmetrical fault current. An analysis of a fault current begins at the moment the normal current stops and the short circuit begins. Figure 14-7 shows the typical wave pattern for one phase of a generator's output. The wave form is symmetrical in that equal values of current are displaced above and below the zero reference. The peak values of current of each wave are equal in a no-fault condition. Figure 14-8 shows the waveform at the time of a fault. Since a large current flows, the amplitude at the moment of the fault is large

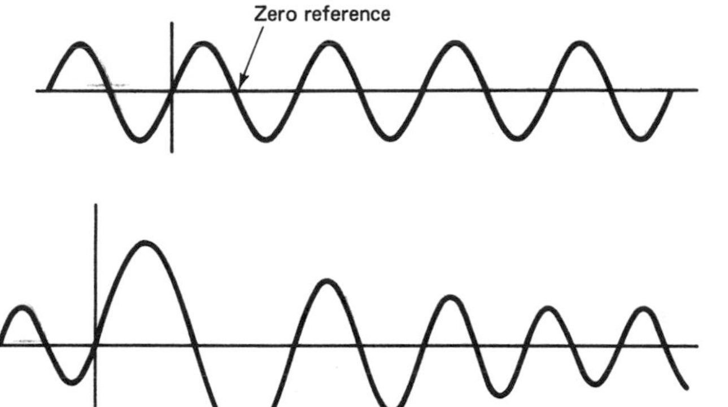

Figure 14-7 Normal symmetrical current.

Figure 14-8 Fault current waveform.

and then it begins to decrease. When a fault occurs, the inductive component of the circuit will cause the reference axis to shift to a value equal to the dc component of the alternating current. This shift, shown in Figure 14-9, will increase the instantaneous peak value of the current to a higher level above the reference axis. In time the waveform will decay to the normal circuit current value. The rate at which it decays to a symmetrical condition is determined by the resistive-inductive components of the circuit. It may take many cycles for the reference axis to return to the zero reference level. Only a few cycles have been drawn for the sake of simplicity. If fault currents were allowed to continue in a circuit during the time that the current values were decaying to the normal state, a large amount of damage to equipment would result. To reduce the damage of the fault due to the time element, it is necessary to interrupt the current during the first few cycles. The time during the first few cycles of the fault is usually termed the *subtransient time*. The time after the first few cycles to the steady state condition is called *transient time*. By using the values of the substransient reactance during the subtransient time and the transient reactance during the transient time, the fault currents can be calculated. Figure 14-10 shows a sample fault calculation for a 25-MVA (million volt ampere) generator with the $X'' = 0.09$ p.u., on a 13.8-

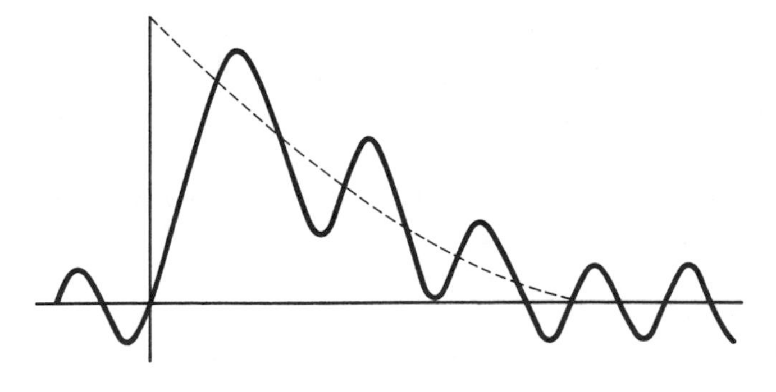

Figure 14-9 DC component of fault current.

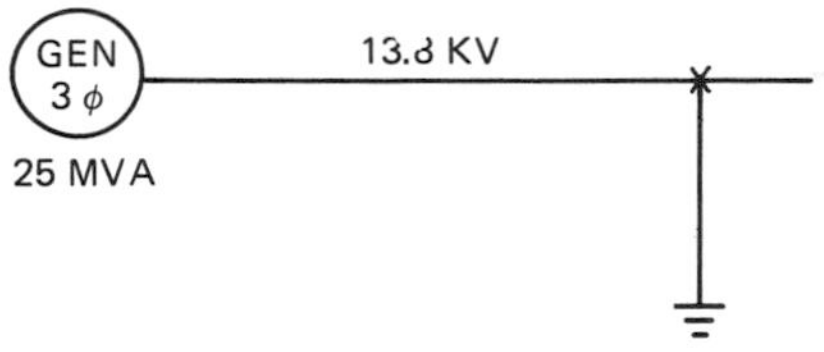

GEN 3 ϕ 25 MVA — 13.8 KV

$$\frac{25 \text{ MVA}}{25 \text{ MVA}} \times 0.09 = 0.09 \text{ p.u.}$$

$$\therefore \text{Fault MVA} = \frac{25 \text{ MVA}}{0.09}$$

$$= 280 \text{ MVA}$$

$$\text{Fault current} = \frac{280 \text{ MVA}}{1.732 \times 13.8 \text{ KV}}$$

$$= 11621.7 \text{ Amps}$$

Figure 14-10 Sample fault current calculation.

KV line. A 25-MVA base will be used. Only the subtransient reactance will be used, since the fault will be at the source. The fault current for the subtransient period is 11,621 amperes. This is a symmetrical current, and it must be multiplied by a correction factor to obtain the asymmetrical value. The correction factor is a function of the reactance to resistance ratio of the circuit. The typical value for a fuse above 600 V is 1.6 and 1.25 below 600 V. The corrected value is then 18,594 A.

Fault Isolation

If the fault current path is broken from the generating source early in the first cycle, then the amount of current will be decreased. Figure 14-11 illustrates the first cycle of the fault current and a typical time when a fuse would open and disconnect the fault from the circuit. The sequence of events are as follows:

1. It can be seen that the maximum value of the current is much less than the possible fault current calculated. This is because there is a fuse in the circuit, and it opens at a low current level compared to the maximum possible level.
2. Due to the high amperage involved, there will be a period of time when an arc will exist in the fuse. The arc will continue until the fuse link has melted to the point that the arc can no longer jump the open gap created by the fuse.
3. The total time for both of these conditions to be completed is known as the total clearing time. The fuse first limits the current and then begins to reduce the time that the current exits as it decays toward zero amperes.

Charts are produced by fuse manufacturers that show the characteristics of fuses during the total clearing time. A chart used to convert symmetrical fault

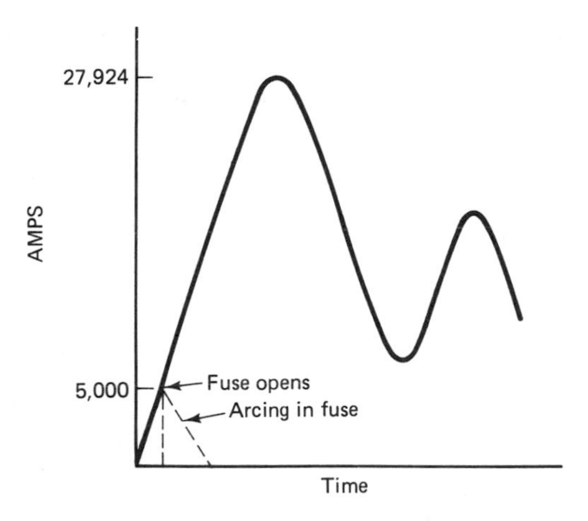

Figure 14-11 Fuse functioning during first cycle.

current values to instantaneous peak values is illustrated in Figure 14-12. Let's examine how to use one of these charts. The current can be derived by entering the bottom of the chart with the symmetrical current value and reading up to the base line and left to the peak value of the current.

1. The value of normal symmetrical current is calculated by Ohm's law, or by current ratings given on the equipment.
2. Starting on the chart at the symmetrical current, move up the chart until reaching the base line.
3. Move left until reaching the edge of the chart and read the value indicated. This is the peak instantaneous current.
4. Base lines can vary to compensate for power factor. Typical multipliers are 2.3 and 2.6. (Figure 14-12 is based on a value of 2.4)

An 11,621-A symmetrical current would have a peak value of 27,934 A. The chart becomes more valuable if the characteristics of fuses are also entered. Figure 14-13 illustrates the chart from Figure 14-12 with the characteristics of fuse A and fuse B superimposed. Re-entering the chart at the 11,621-A fault current and reading up to the characteristic of fuse A and left to the base line, the operating point of the fuse is located. By knowing the operating point, two important values can now be derived.

1. By continuing left from the operating point, the peak value of the current when the fuse begins to function is found. For the 11,621-A fault current this value would be 5,000 A.
2. By reading down from the operating point to the symmetrical current line,

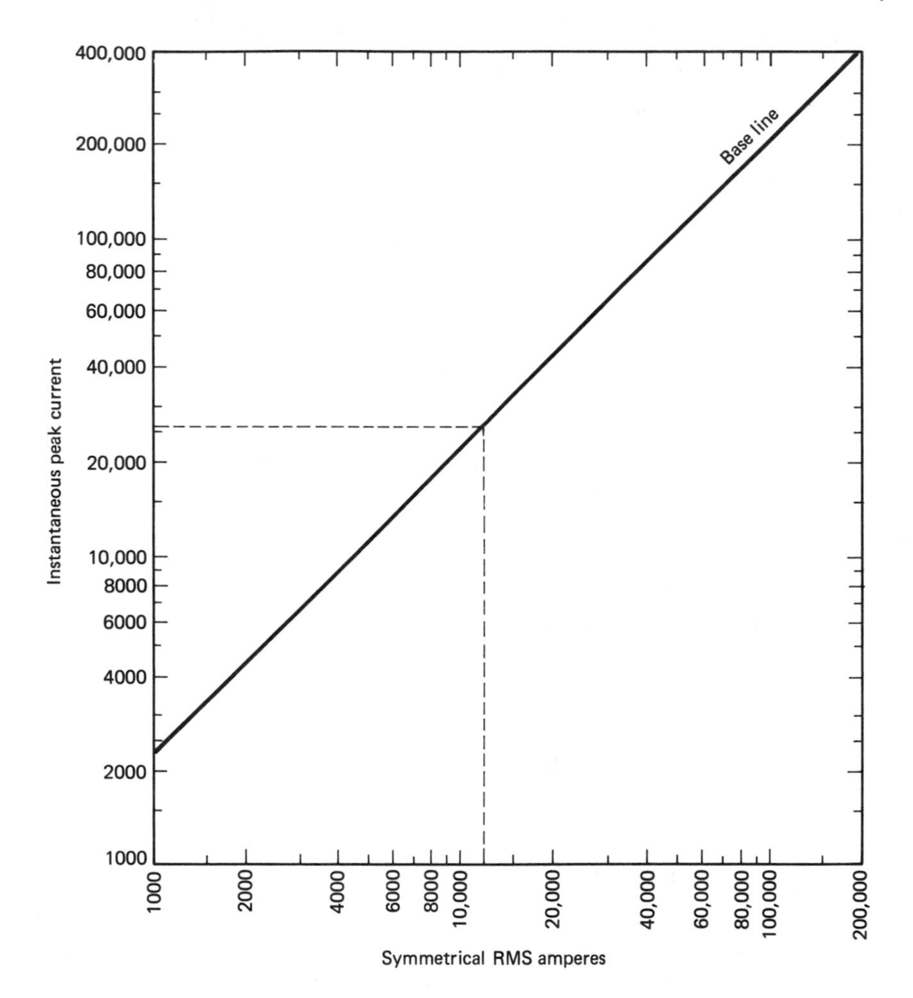

Figure 14-12 Instantaneous current calculations

the actual symmetrical current value when the fuse begins to function is found. In this example the current would be 2,200 A.

The 11,621-A fault had the potential to reach a peak value of 27,924 amperes but was limited to 5,000 A. The symmetrical current was limited to 2,200 A. This illustrates the protection provided by a fuse, and with similar charts, circuit breakers.

Time Current Curves

Fuses have many different design characteristics. The time that it takes for a fuse to open is determined by the physical construction of the fuse. Time-current

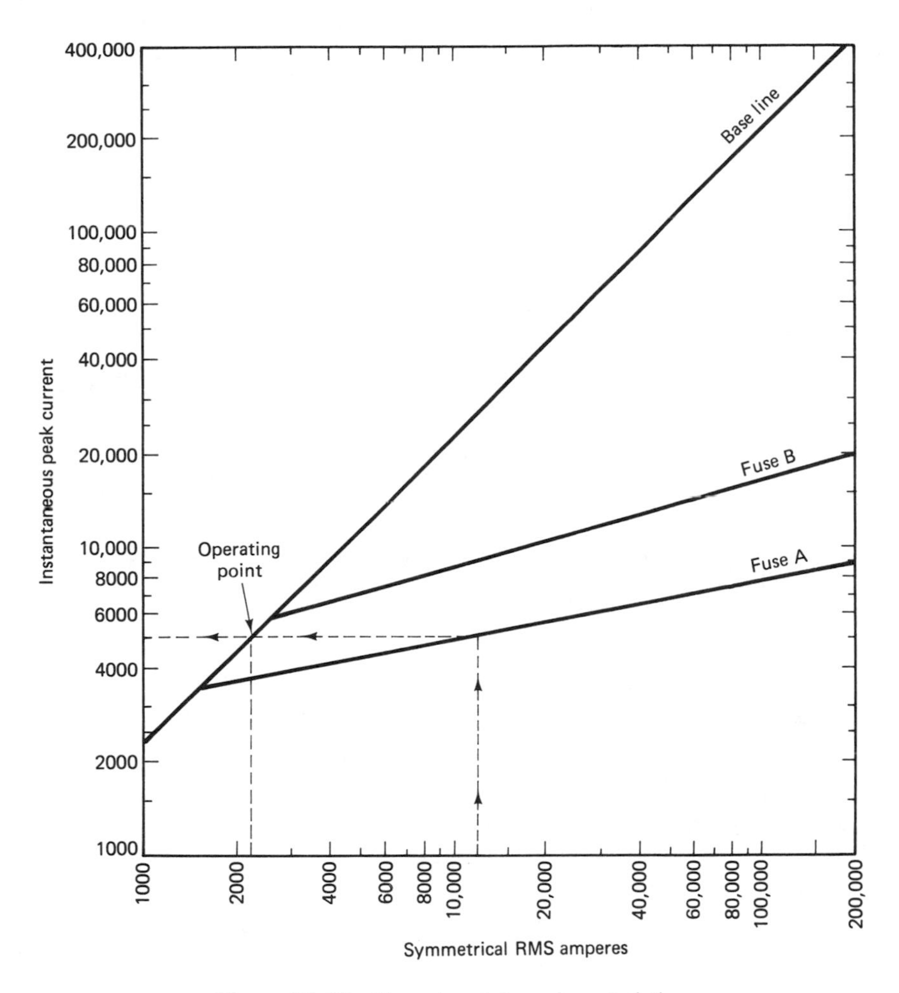

Figure 14-13 Current and fuse characteristics.

characteristic curves are available from fuse manufacturers which show the relationship of the current in the fuse to the time of functioning. There are two characteristics plotted using time-current characteristics. Minimum melt time and maximum clearing time. Figure 14-14 illustrates the typical curve for the minimum melt time of two fuses. By reading the fault current value on the bottom of the chart up to the curve which corresponds to the fuse in use and then left to the time line, the minimum melt time can be calculated. Figure 14-15 illustrates the typical curve for the maximum clearing time. The same procedure is used to calculate the maximum clearing.

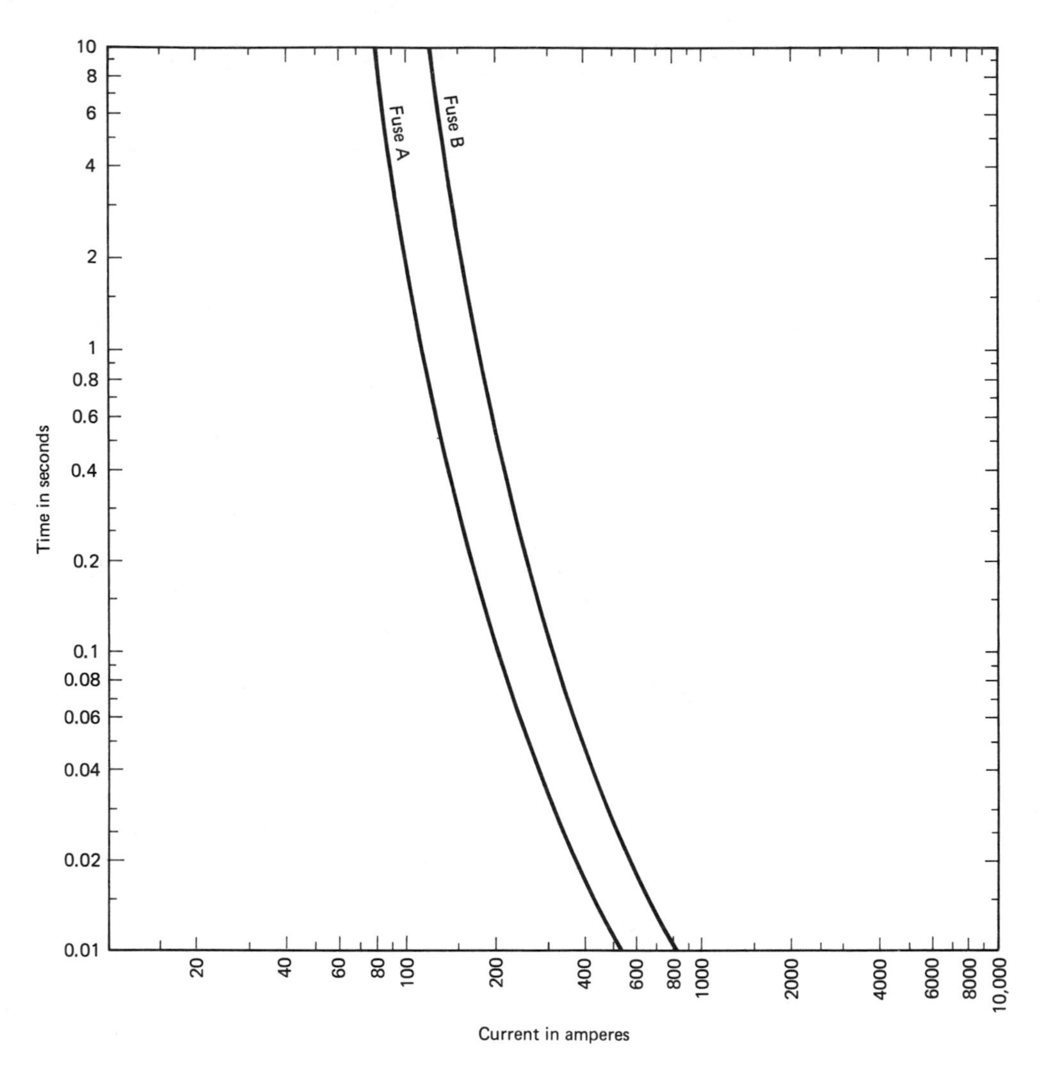

Figure 14-14 Minimum clearing time.

Fuse Coordination

When a fault occurs, the fuse nearest the fault is the one that needs to isolate it from the energy source. If a fault jumps a fuse and is cleared by the next fuse, then unnecessary blackouts and shutdowns will occur in a distribution system. Fuse jumping is usually the result of improper system design or improper fuse replacement. The corrective action for this problem is called *fuse coordination*.

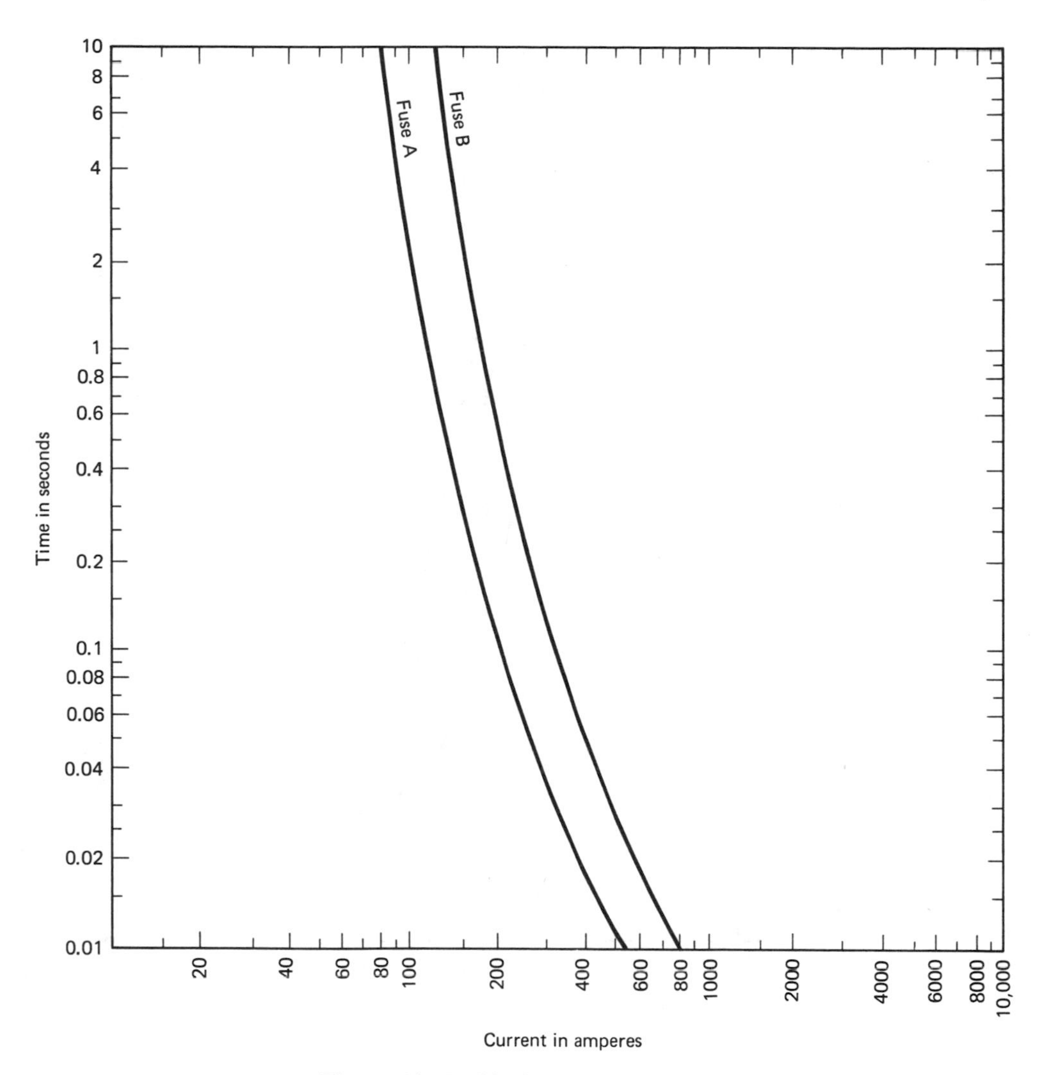

Figure 14-15 Maximum clearing time.

By superimposing the melting and clearing time characteristics on one chart, for example, as shown in Figure 14-16, the functioning times of the fuses can be coordinated. Using the fault current value and reading up to the fuse in use and then left to the appropriate time line, the time for the fuse to melt or clear can be calculated. Proper coordination dictates that the fuse clearing a fault must function before the minimum melting time of the next fuse in the system. Assuming a fault current of 200 amperes, the maximum clearing time and minimum melting times of fuse A and fuse B can be compared.

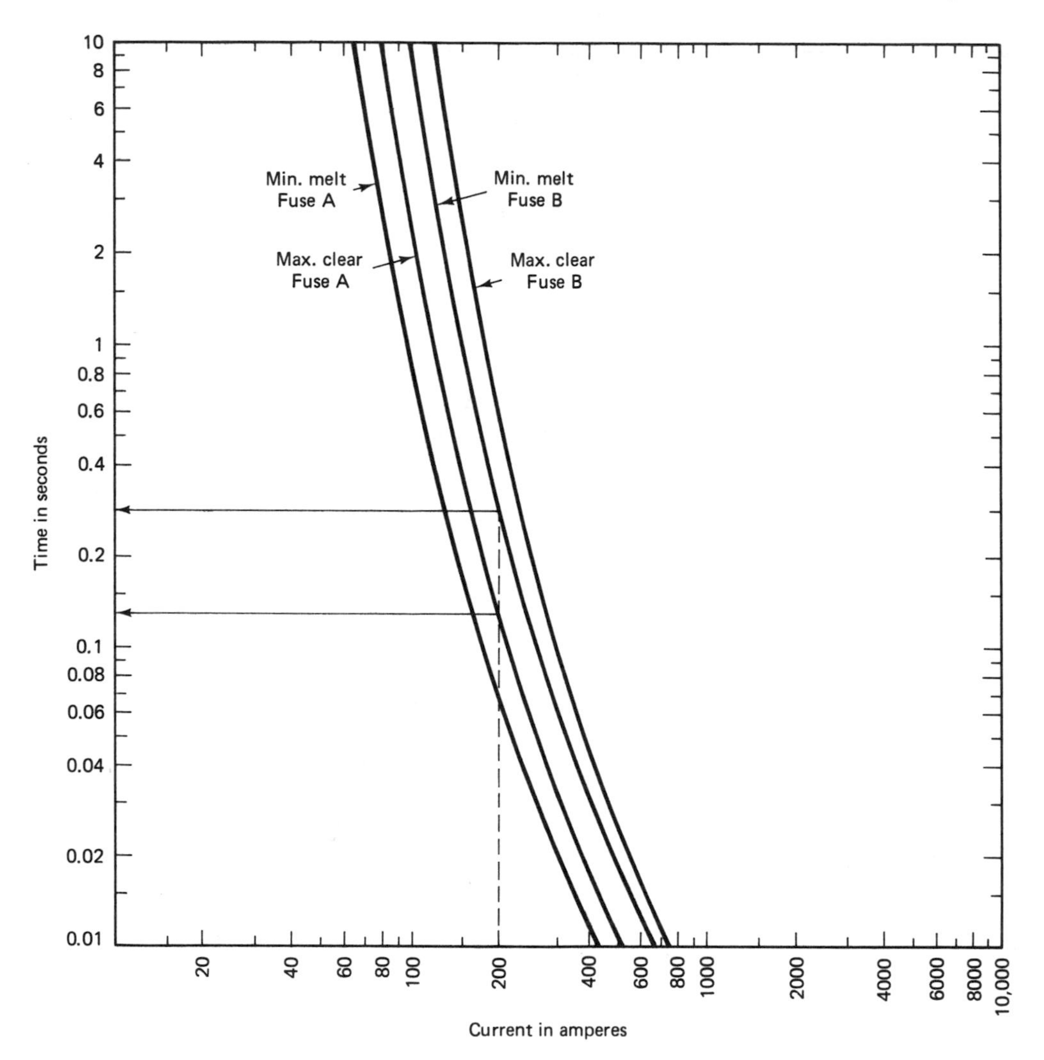

Figure 14-16 Fuse coordination.

1. Fuse A has a clearing time of 0.13 second.
2. Fuse B has a melting time of 0.28 second.
3. Fuse A will then clear a fault before fuse B and thus fuse jumping will not occur.

Typical ratios of 2:1 and greater are used when possible as an additional precaution to prevent needless blackouts. Check with the supplier of your fuses for data sheets. Proper design can not only protect your equipment but can also avoid shutdowns during production runs.

SUMMARY

Protection of the energy supply system is the ultimate purpose of a fuse. Proper fault current protection goes beyond installing a 30-ampere fuse for a 24-ampere load. The current interrupting capacity of the fuse is a very important consideration if fault current damage and needless blackouts are to be prevented. To select a fuse properly, the entire energy system is a source of current. Coordination of each branch will assure that fault current values will be limited and isolated before they can damage associated equipment.

EXERCISES

On a separate sheet of paper, complete a response for each question, statement, or problem listed below.

14.1. Name one short-circuit interrupting device.
14.2. How does a fuse prevent excessive current from going through it?
14.3. Is is normal to replace a 5A fuse with a 10A fuse? Under what condition would it be proper to replace a fuse with one that has a higher rating?
14.4. What might indicate a short-circuit problem in a circuit?
14.5. List the two most important functions of a fuse.
14.6. Describe what circuit breakers are.
14.7. In what way does the function of a circuit breaker differ from that of a fuse?
14.8. Describe what is meant by *fault* current.
14.9. What determines the magnitude of fault current damage?
14.10. List the device that limits the duration of a fault.
14.11. Identify the three kinds of current used in fault analysis.
14.12. Describe an effective method for isolating a fault.
14.13. A generator's impedance is composed of what two things?
14.14. What parameter causes the line impedance to change?
14.15. What does *per unit value* mean?
14.16. What determines the amount of time that it takes for a fuse to open?

15

Industrial Motors

OBJECTIVES

Upon successful completion of this chapter, you should be able to

(1) differentiate between direct current and alternating current motors
(2) explain counter-electromotive force and magnetic interaction and describe how they cause motor rotation
(3) explain the operation of a series-wound DC motor and the precautions that should be followed to ensure its safe operation
(4) describe the shunt-wound DC motor and its important features
(5) discuss the operation and characteristics of a compound-wound DC motor
(6) explain how rotating magnetic fields are formed in an AC motor
(7) describe the operation of single and split-phase motors
(8) describe a capacitor-start motor and discuss the related troubleshooting procedures
(9) explain how a poly-phase motor operates and describe appropriate troubleshooting procedures

INTRODUCTION

A large portion of all control circuits are designed for the purpose of controlling electric motors. The motors may vary in size from that of a small battery-powered unit to a several-hundred horsepower motor. The common measure used to define the size of a motor is the horsepower. In electrical terms, one horsepower is equal to 746 watts. In mechanical terms it represents the amount of force required to move 33,000 foot-pounds in one minute. The physical size of the motor increases as horsepower increases. Motor voltage also is a factor in the size of the motor. By using large voltage sources, the physical size can be reduced.

Two main motor categories are direct-current and alternating-current. This chapter will examine the basic uses and circuits of each as they relate to a control circuit. As a general rule, AC motors are easier to hook up and require fewer components in the circuit, but they are more difficult to control for precision movements. Direct-current motors, on the other hand, need many more components but allow much more precise control of the motor's movements. A one horsepower motor will do the same amount of work regardless of whether it is run by an AC or DC source. It is the degree of control that we desire that makes one more useful than the other. Advances in speed control circuits are making control of AC motors more feasible. However, this still does not always allow the degree of precision that direct-current motors can provide.

DIRECT CURRENT MOTORS

As we have emphasized, precise control is one of the major features of the direct-current motor. Several other characteristics must also be considered in choosing these motors. These characteristics are a function of how the motor is constructed and the wiring configuration used. Three typical configurations will be examined: the series-wound, shunt-wound and compound-wound motors. Speed control, torque developed, and horsepower available are determined by the configuration that is used, and this is sometimes the determining factor in choosing one wiring system over another. Essentially, all motors are physically the same as generators. The exception is that instead of generating a current from mechanical energy the motor uses current to generate mechanical energy.

Counterelectromotive Force

By introducing a current into a motor coil, a magnetic field is created. The magnetic interaction of the coil's magnetic field with the motor's magnetic field causes the coil to rotate. The rotation of the coil is the motoring action. This is shown in Figure 15-1a. One point that is important to realize is that this rotating coil also works in the same manner as a generator. This rotating coil generates a current of its own, as illustrated in Figure 15-1b. The polarity of this current opposes the current that is causing the rotation. The resistance of the wire combined with the effect of the currents develops two opposing voltage drops in the circuit. The voltage that is used to run the motor is called the applied voltage, or applied emf (electromotive force). The voltage created by the generator action of the coil is called the counter electromotive force (cemf), since it is opposing the applied emf. The resulting voltage is actual voltage present in the motor due to the interaction of the two opposing voltages. The actual voltage is always smaller than the applied voltage, and this also reduces the amount of current that exits in the winding. Figure 15-1c illustrates the effect of the cemf. This is important because it permits using smaller wire sizes for motors and for use in control circuits. This results in a savings when installing the equipment. The same principle is also applied to AC motors.

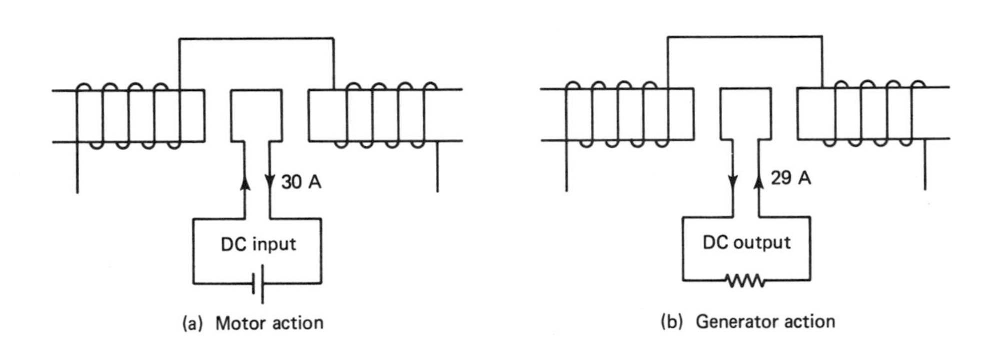

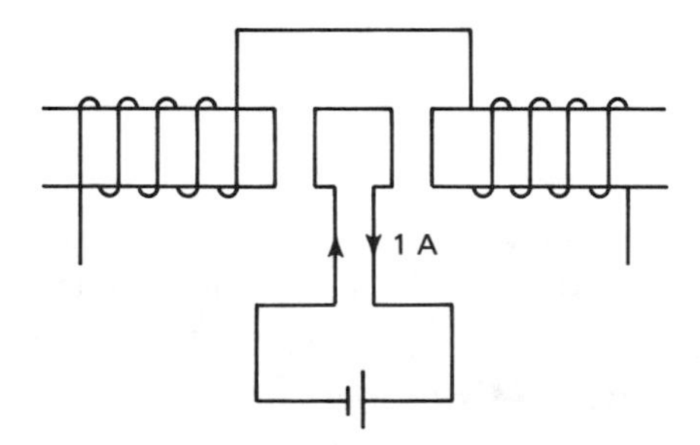

(c) Effect of CEMF with motor action

Figure 15-1 Effect of CEMF.

Series-wound DC Motor

A series-wound motor circuit diagram is shown in Figure 15-2. The circuit gets its name from the fact that the field coil is placed in series with the motor armature. Let's evaluate its operation.

1. Current is supplied by the DC source and flows through the field coil and armature equally since they are in series.
2. Assuming a constant load on the motor, this current will remain at a constant level. Cemf is stabilized at a set level.
3. The load decreases.
4. The inertia of the motor lets it accelerate, and this causes the cemf to increase.
5. This causes a decrease in armature current.
6. Decreasing current in the armature causes a decrease in torque. The armature begins to decrease its rate of acceleration until it stabilizes at a new level above the initial running speed.
7. Simultaneously, the current is also decreasing in the field winding. This causes the strength of the magnetic field to decrease which also decreases torque.
8. The generator action of the motor turning in the decreased magnetic field decreases the effect of the generator voltage developed, cemf.
9. The decreased cemf opposes the applied voltage less and causes the running voltage to increase.
10. The torque required to maintain the running speed is now stabilized and at the minimum required level to maintain this speed.
11. If the load were to increase, the armature speed would decrease.
12. The decrease in the armature speed reduces the cemf developed.
13. The running voltage increases, and the current drawn by the motor also increases.
14. As the current increases, the torque developed also increases to maintain motor speed.

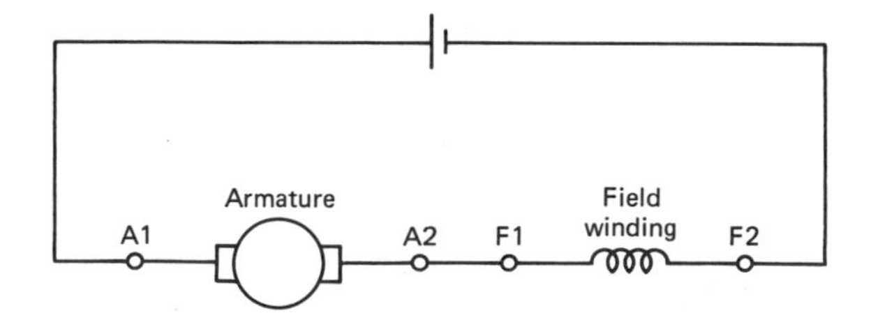

Figure 15-2 Series-wound DC motor.

A special problem can exist with the series-wound motor. Running the motor without a load can cause it to continue to increase speed to the point that it can destroy itself with centripetal force. The load is therefore a very important current-limiting factor. If there is any condition which may result in the motor's having an unloaded condition, it should not be used. Gear drives are therefore preferable over belt drives.

Typical applications include drag lines or overhead cranes that must lift or pull heavy loads that will remain connected while being moved. Figure 15-3 illustrates the relationship of torque and armature current in a series-wound motor while Figure 15-4 shows the relationship of armature current and motor speed. Increasing armature current will cause an increase in torque but will also cause a decrease in motor speed. Several facts can be summarized about the series-wound motor.

1. Starting torque is high due to limited cemf.
2. Running torque is low.
3. Motor speed changes with load changes.
4. The series motor must always have a load connected.
5. Changing rotation is accomplished by reversing either the field or the armature connections but not both at the same time.

Shunt-wound DC Motor

The schematic of the shunt-wound motor is shown in Figure 15-5. This configuration gets its name from the fact that the field shunts the armature. Let's examine the operation of this motor and then summarize its characteristics.

1. First, remember that the field coil and the armature are both in parallel and therefore will have the same voltage across them in this circuit.
2. Assuming the motor is running at a constant load and speed, the currents in each branch of the circuit will be stabilized.
3. Increasing the load will cause the motor speed to decrease.
4. Decreased speed reduces the cemf developed and increases the amount of current flowing through the armature.
5. Increasing the torque causes the torque to increase, and the motor speeds up.
6. The field coil voltage remains the same as the line voltage and thus provides a constant magnetic field for the armature.
7. With a constant magnetic field, decreased motor speed decreases the cemf.

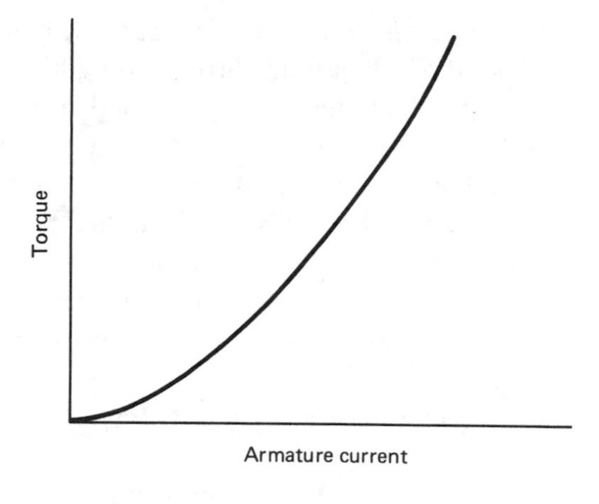

Figure 15-3 Series-wound motor torque vs. armature current.

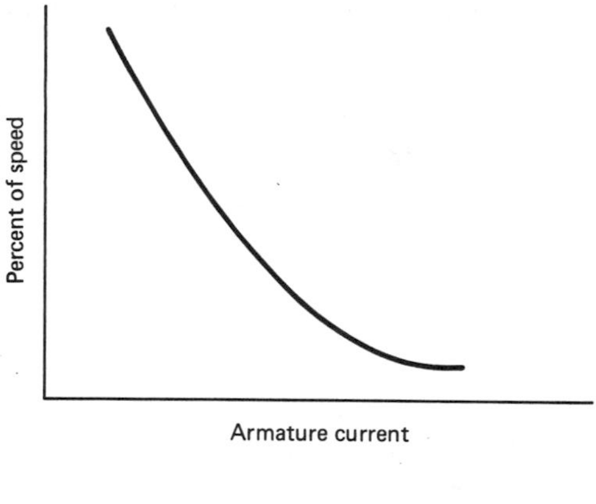

Figure 15-4 Series-wound motor speed vs. armature current.

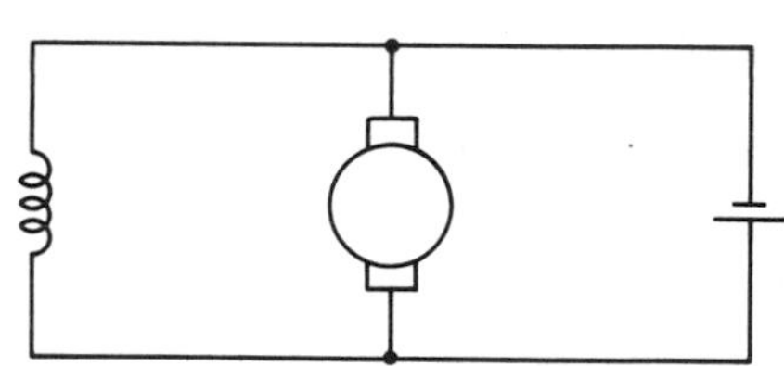

Figure 15-5 Shunt-wound motor.

8. Decreasing the load lets the motor speed up, and this increases the cemf developed.
9. Armature current then will decrease along with the torque.
10. The decreased torque will tend to slow the motor speed.

Many of these same principles are the same as those for the series-wound motor except the series-wound motor has a few large windings in the field and the shunt-wound motor has many small windings in the parallel field. The series-wound motors allow the field winding current to fluctuate with the armature current while the shunt-wound has a constant field current strength independent of the armature current. Figure 15-6 illustrates the torque and armature current characteristics while Figure 15-7 illustrates the speed and armature current relationships for a shunt-wound motor. Applications include loads that vary in magnitude but still must run at a constant speed. Since the field coil helps determine the amount of cemf developed, it is necessary to ensure that the motor does not run without the field coil connected or open. An open field coil will allow cemf to decrease to a low level, and the motor will run away or burn up due to high currents and high speeds.

Several other important characteristics are listed below.

1. The shunt-wound motor has a constant speed with load changes.
2. Starting and operating torques are good but not as high as those of the series-wound motor.
3. Shunt-wound motor rotation is changed by reversing either the field or armature current windings, but not both.

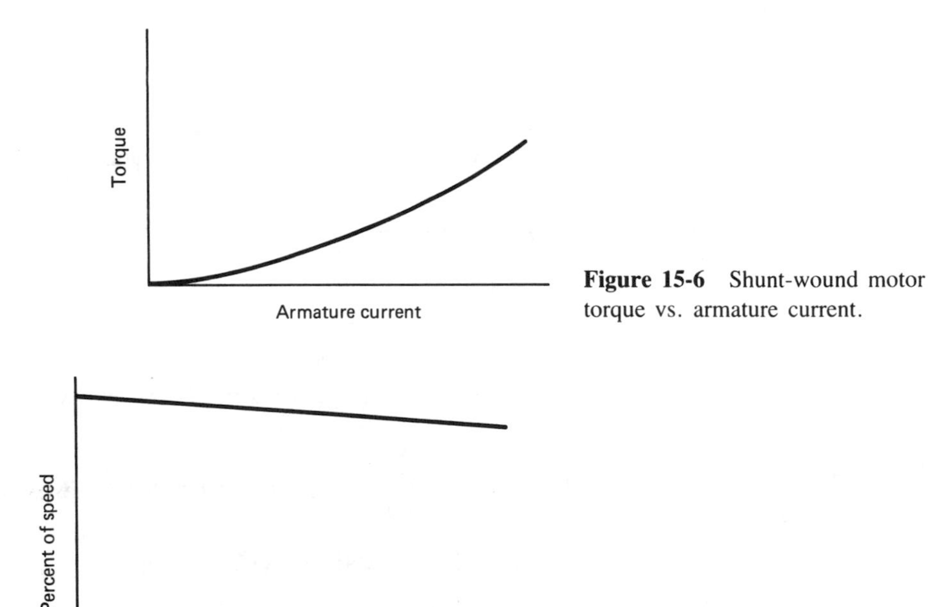

Figure 15-6 Shunt-wound motor torque vs. armature current.

Figure 15-7 Shunt-wound motor speed vs. armature current.

Compound-wound DC motor

A third type of DC motor is the compound-wound motor. It has two field coils instead of one. It combines the benefits of both the series-wound and shunt-wound motors. Two configurations are possible: the cumulative compound and differential compound. The cumulative-compound version is used more frequently, so we will limit our discussion to its configuration.

As illustrated in Figure 15-8, two possible arrangements are possible with this version. Depending on the polarity of the field coil and its placement in the circuit, it may be designated a long-shunt or short-shunt configuration. The long-shunt configuration has the shunt-field (high resistance coil) in parallel with the armature and series field. In the short-shunt configuration, the shunt field is in parallel with the armature, but both are placed in series with the series field coil.

The low resistance value of the series coil has minimal effect upon its location in the circuit. The operating characteristics therefore remain nearly the same for both configurations. The sequence of operations is the combination of both the series and shunt motors previously discussed. Applications are similar to that of the shunt motor except it tolerates rapidly changing loads better. The torque and speed relationships with changing armature currents are given in Figures 15-9 and 15-10. A summary of the characteristics include:

1. Good speed regulation.
2. Higher starting torque than shunt motor.
3. Runaway protection is not required.

Speed control is one of the important advantages of a DC motor (methods of speed control will be examined in a later chapter). A secondary advantage of speed control is to allow position control. Starting and controlling is also a function of speed control.

Several important formulas may be helpful for analysis of DC motors. One is the speed of rotation. The constant C is a design feature and for basic analysis may be ignored.

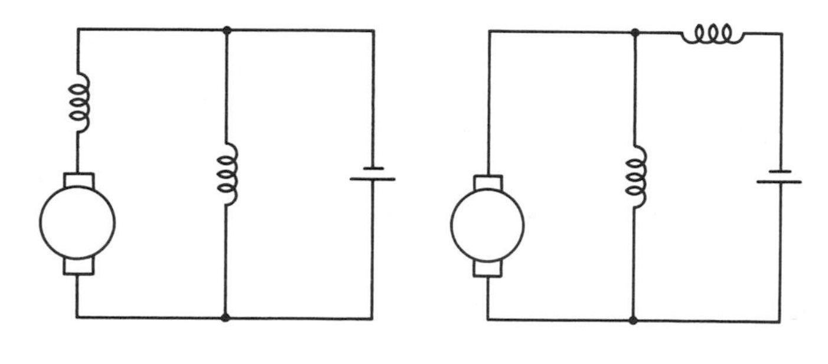

Figure 15-8 Compound-wound DC motor.

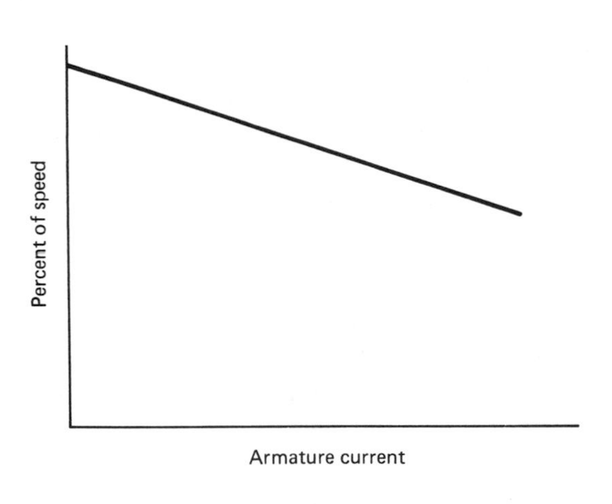

Figure 15-9 Compound motor speed vs. armature current.

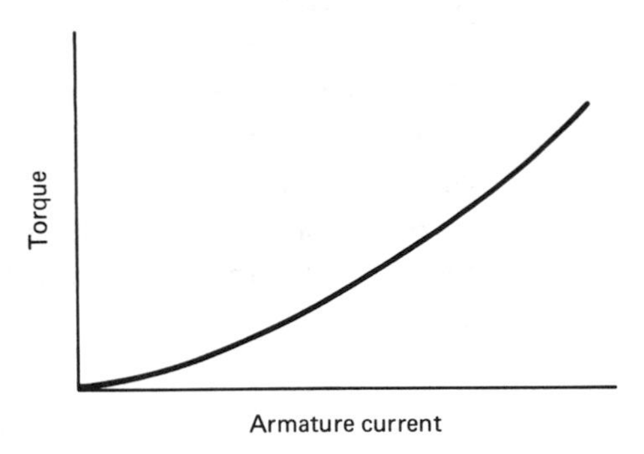

Figure 15-10 Compound motor torque vs. armature current.

$$\text{Speed of rotation} = \frac{C_{\text{emf}}}{\text{Field strength} \times C}$$

A second formula for evaluating the torque of the motor may also be useful. It is not used to calculate an exact value but for overall evaluation only.

$$\text{Torque} = \text{Field strength} \times \text{Armature current}$$

AC MOTORS

When speed or precise control is not a major factor in selecting the motor, many times the AC motor will be more economical. Several different types of AC motors are made, and again each has some advantages and disadvantages. Speed control and rotation direction changes are possible, but they generally require more control equipment than DC motors (with the exception of three-phase AC motors).

Rotating Magnetic Fields

Overlooking all other purposes and advantages of AC motors, the single most important principle that should be understood is that of the rotating magnetic field—since this is what makes all motors work. Figure 15-11 shows a simple set of poles with their windings. When an AC signal is applied to the poles during times zero to two, they have a polarity as shown in the diagram. Of course, the intensity of the magnetic field varies with the amplitude of the signal. In a similar manner, as shown in Figure 15-12, during the second half cycle, the polarity will be reversed. This simple motor is not very effective. To improve it, more poles are added, as illustrated in Figure 15-13. The polarities shown are for the first half cycle at time one. For the second half cycle, the polarities would be reversed. Notice that there are two generators in this diagram. This causes a ninety-degree phase difference between the two poles, which creates a magnetic difference in potential in the same manner as a voltage potential difference is in an electrical circuit. Figure 15-14 illustrates the polarities developed at each pole during each time indicated. Follow point X at each period in time to see how it rotates around the motor. If a rotor is placed in the motor with a polarity opposite this rotating field, it will follow it around, producing motor action.

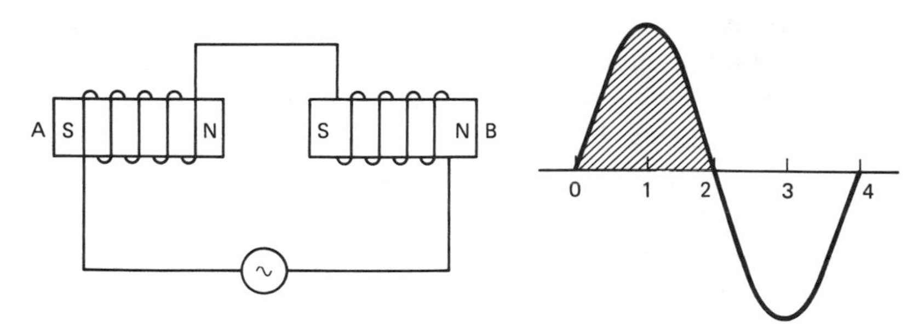

Figure 15-11 Simple motor, first alternation.

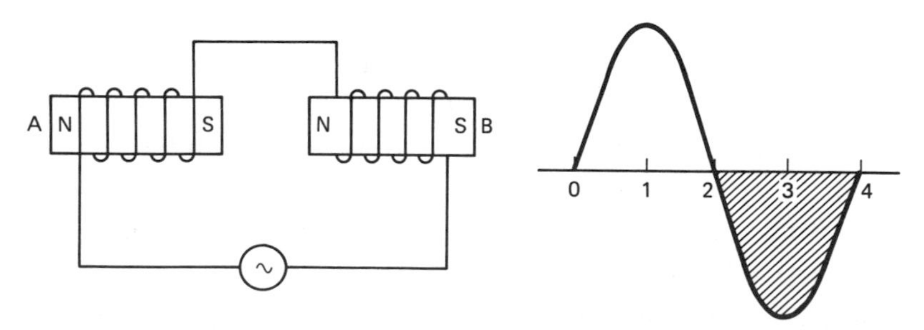

Figure 15-12 Simple motor, second alternation.

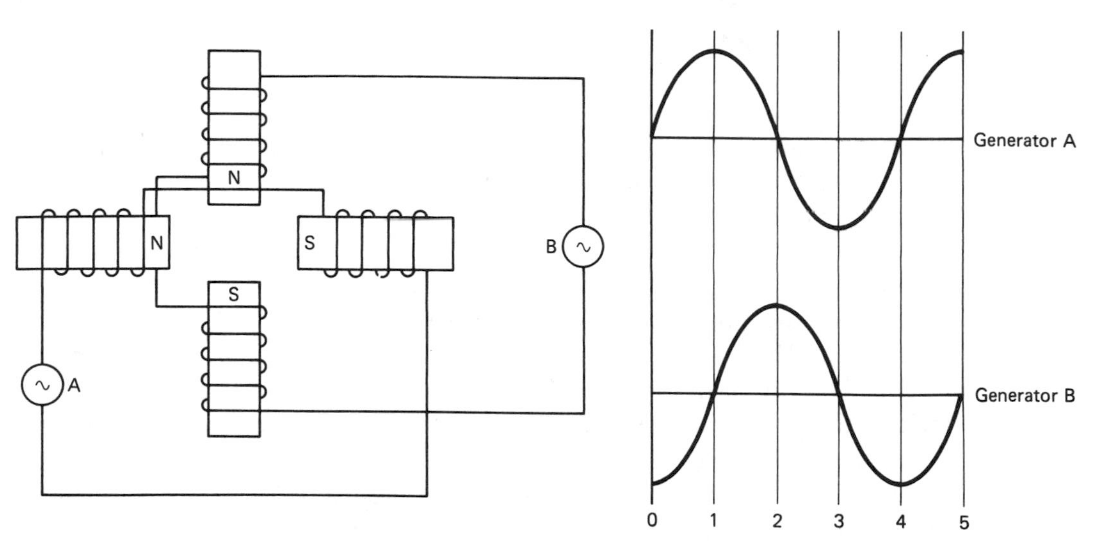

Figure 15-13 Refined motor.

Single or Split Phase Motors

The first type of AC motor we will examine is the single phase motor. Figure 15-15 shows a circuit diagram and simplified motor diagram. The phase shift required for rotation is accomplished by the physical arrangement of the windings and the difference in inductive reactance of the two windings. The rotating magnetic field that is created and the interaction of the polarized rotor causes it to rotate. The number of poles per phase helps to increase the effect of the rotating magnetic field, and this results in a speed change for each motor. As the diagram indicates there are two separate windings. One is a start winding, and the other is the run winding. Let's examine the steps in the normal operation.

1. On startup the start and run windings are in parallel with the line voltage source.
2. The physical ninety degree placement of the coils with the inductive reactance starts the rotor turning.
3. Each separate pole helps the rotor to turn a little further each time.
4. After the speed of the motor has increased, the inertia of the rotor keeps it running with just a slight push from each pole as the rotor moves past it.
5. At this time the start winding is no longer required to maintain rotation.
6. The centripetal switch opens the start winding circuit, and only the run winding remains in the circuit.

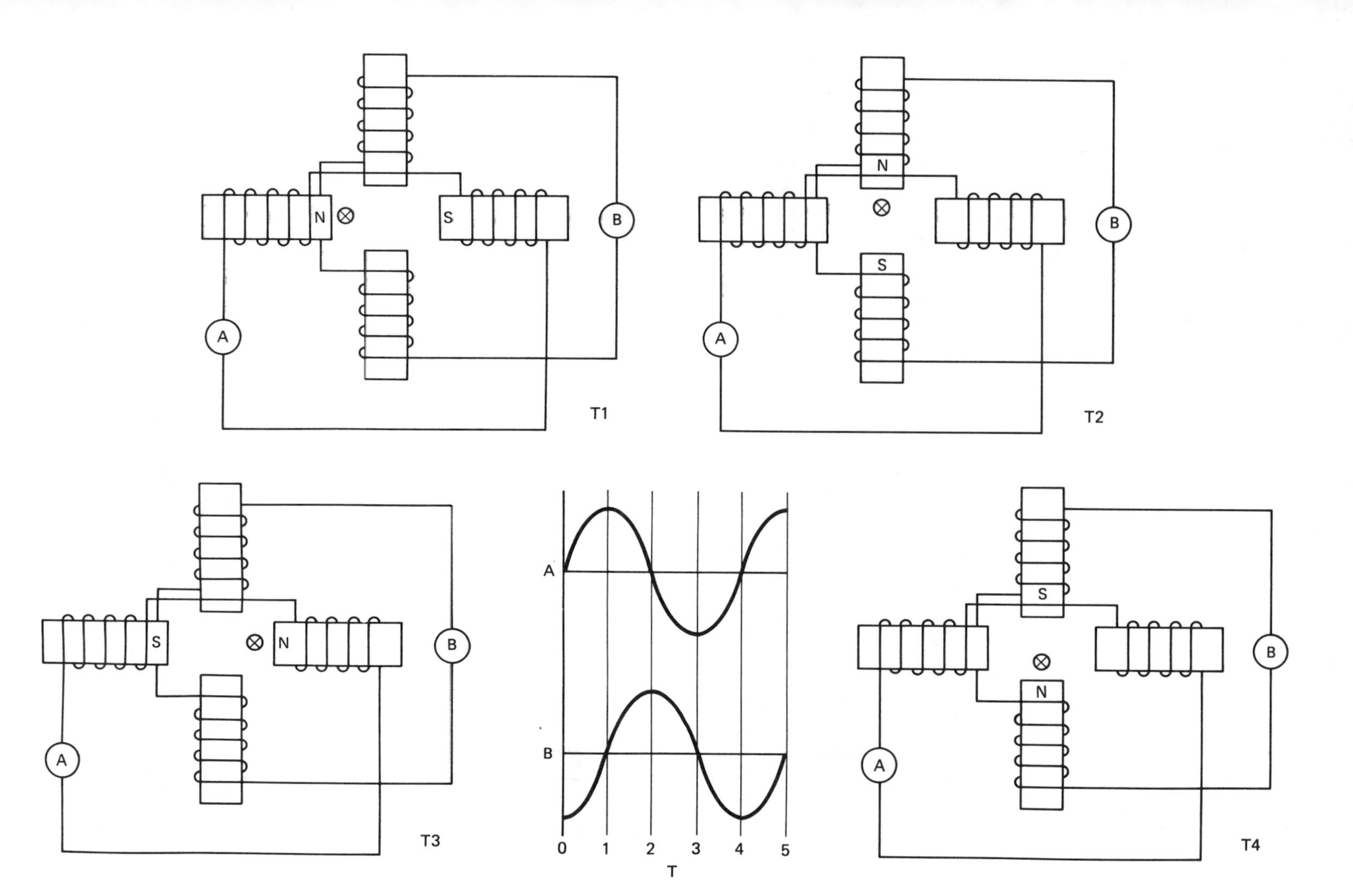

Figure 15-14 Rotating magnetic field.

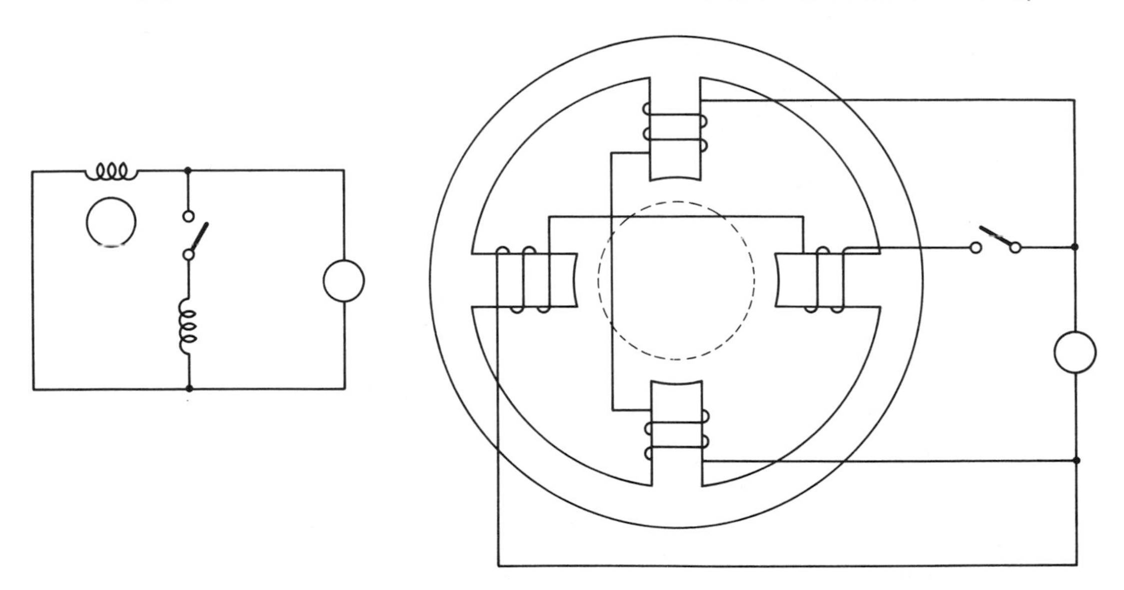

Figure 15-15 Split phase motor.

Capacitor Motors

The torque created by a split-phase motor is not always sufficient to start turning a load. This is partially due to the method of construction used for this type of motor. By including a capacitor, a single-phase source can be used, and a near ninety-degree phase relationship can be developed without using two separate generator sources. Figure 15-16 gives an example of a typical capacitor motor. The only electrical difference in this circuit from the split-phase motor is the addition of a capacitor. A centripetal switch opens in the run winding in a similar manner.

Troubleshooting Capacitor Motors

Several problems can occur in a capacitor-type motor. Figure 15-17 shows a common nameplate diagram used for changing rotation and for changing voltage connections. Refer to this diagram as needed.

1. Check for power to the circuit. This may seem rather simple, but you will be amazed at how many times this cures the problem.
2. Check for continuity at leads T1 and T2. A low resistance reading indicates a complete winding. A high reading indicates an open or burnt winding.

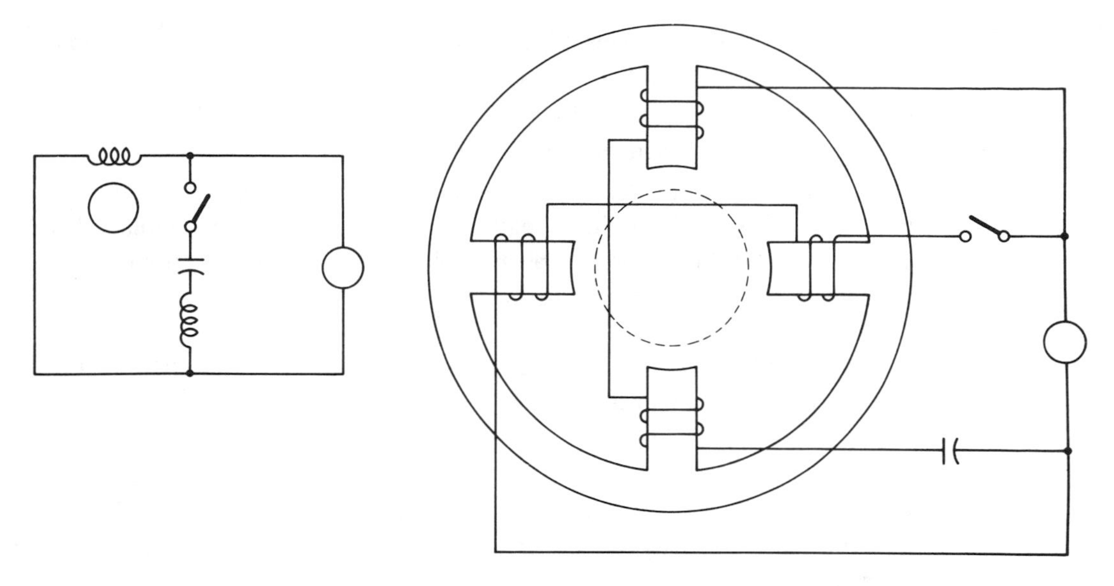

Figure 15-16 Capacitor motor.

3. Check for continuity at leads T3 and T4. Follow the same procedure as in step two.
4. Disconnect the capacitor and short the terminals together with a screwdriver. Large voltages may be stored in capacitors, and this will discharge them.
5. Connect the leads from the capacitor together, T6 and T7.

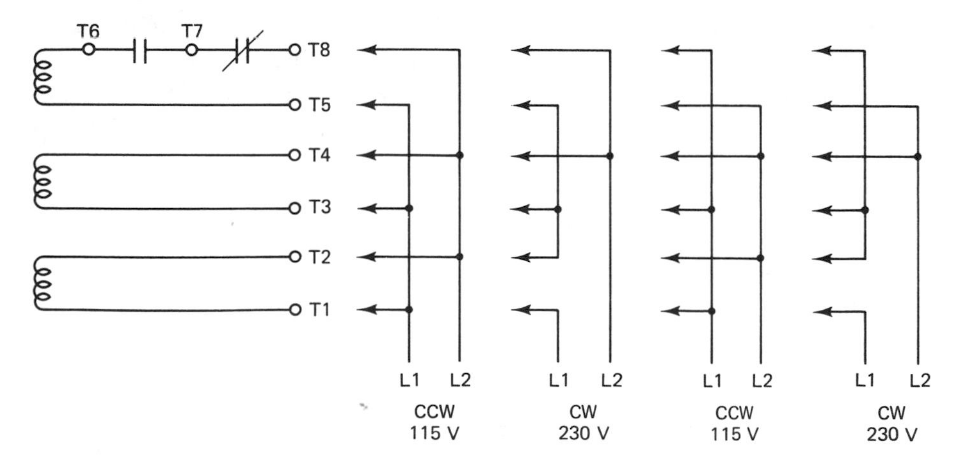

Figure 15-17 Capacitor motor nameplate diagram.

6. Check for continuity at leads T5 and T8. A low reading indicates a complete circuit through the winding and centripetal switch.
7. Reconnect all leads and try a new capacitor if available. A testing circuit is given in Figure 15-18 to test a capacitor.
8. Place a clamp-on ammeter on the line connection, and check the amperage during motor startup. A high reading that drops to the nameplate level indicates a bad capacitor has been replaced. A continuous high reading indicates a possible bad centripetal switch.

It may be desirable to test the capacitor to see if it has retained its rated value. The circuit given will allow you to do this.

1. Construct the circuit.
2. Turn on the power to the circuit and note the condition of the fuse. A blown fuse indicates a shorted capacitor. Replace the capacitor and fuse.
3. Note the readings of both the ammeter and voltmeter.
4. Calculate the rating of the capacitor using the formula in step 5.
5. Number of microfarads = 2650 × ammeter reading (amperes) ÷ by the voltmeter reading (volts).
6. A reading exceeding five percent of the rating on the capacitor should be replaced. Note the voltage rating of the capacitor before replacing it.

POLYPHASE MOTORS

Large industrial plants use three-phase power because it is more economical than single-phase power. The voltages generated by three-phase units are displaced 120 degrees apart, as shown in Figure 15-19. When applied to a motor, this allows simpler construction. The format is like that in Figure 15-13 except a third

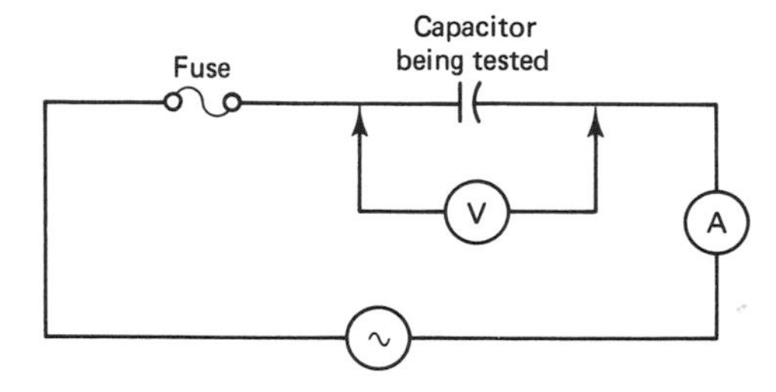

Figure 15-18 Capacitor testing circuit.

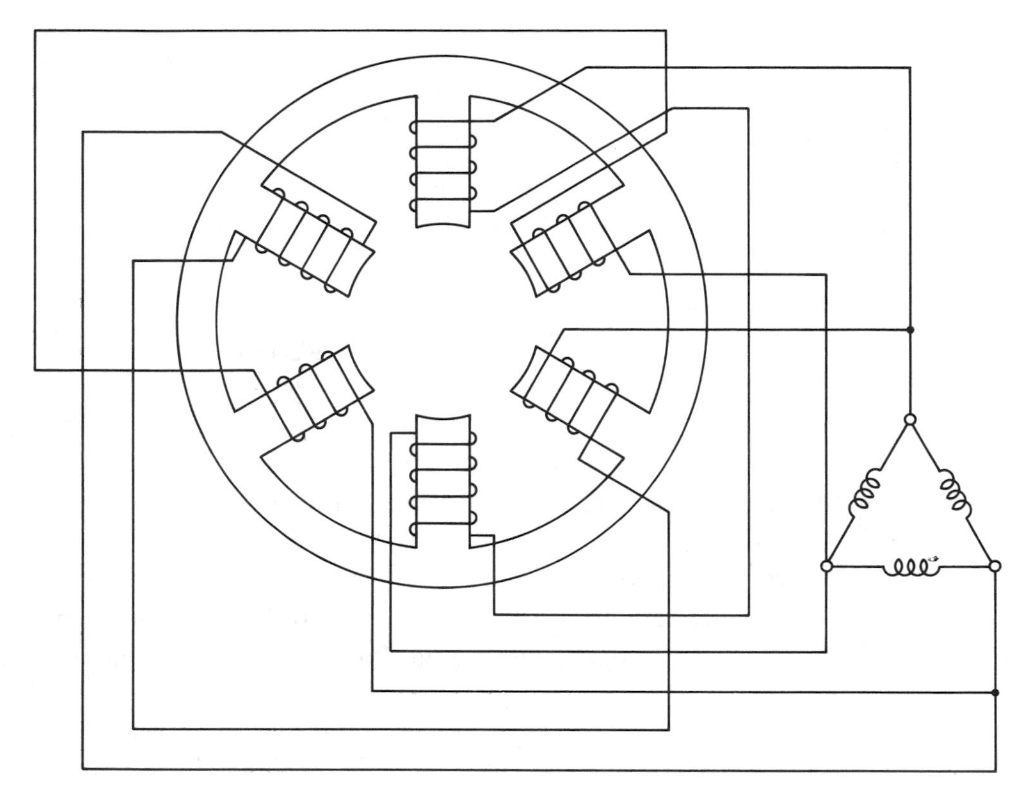

Figure 15-19 Three-phase power.

set of poles is added along with a third generator. Just as for generators, several configurations are found in three-phase motors. Three-phase motors can be wired in a delta or wye connection. The connections made depend on the number of leads brought out of the motor. This varies from six to twelve leads. Figures 15-20, 15-21, and 15-22 show wye and delta connections for six, nine, and twelve-lead motors.

Troubleshooting Three-phase Motors

Three-phase motors are much easier to test than other types of motors. Refer to Figures 15-20, 15-21, and 15-22 or the nameplate on the motor for the diagram of the motor you are testing.

1. Remove the power source from the motor.
2. Using an ohmmeter, measure the resistance of the motor where the line voltage connectors are attached. This can be done at the motor starter some distance away from the motor if necessary.

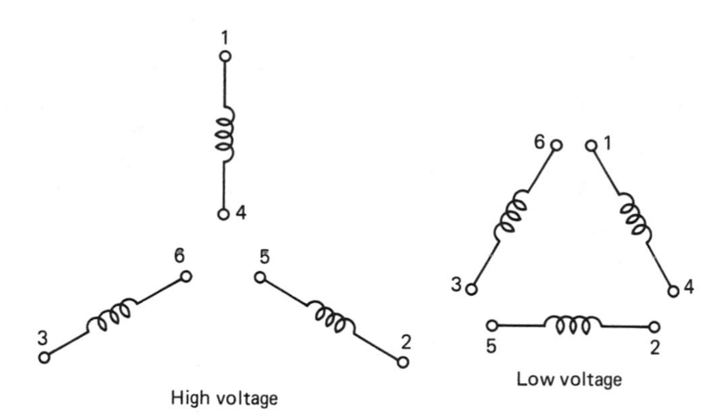

Figure 15-20 Six-lead motor connections.

3. Compare the readings from line 1 to 2, 2 to 3, and 1 to 3.
4. An equal low reading of 3 to 10 ohms between each phase is a fairly good indication that the windings are not open.
5. A zero reading indicates that a short circuit exists in a winding.
6. A resistance reading should also be made to the frame of the motor.
7. A zero reading would indicate a burned or open winding that is shorted to the frame.
8. If a megohmmeter (megger) is available, check the motor again. The megger will give a better indication of the condition of the insulation inside the motor.

These are quick tests for locating faults in motor circuits. The better test involves using the megger. The best method to test a motor fully is to place it on a

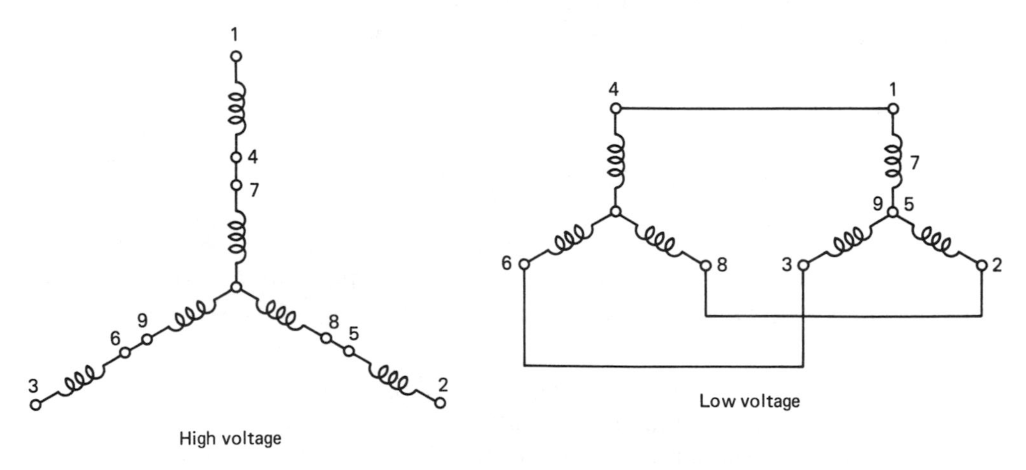

Figure 15-21 Nine-lead motor connections.

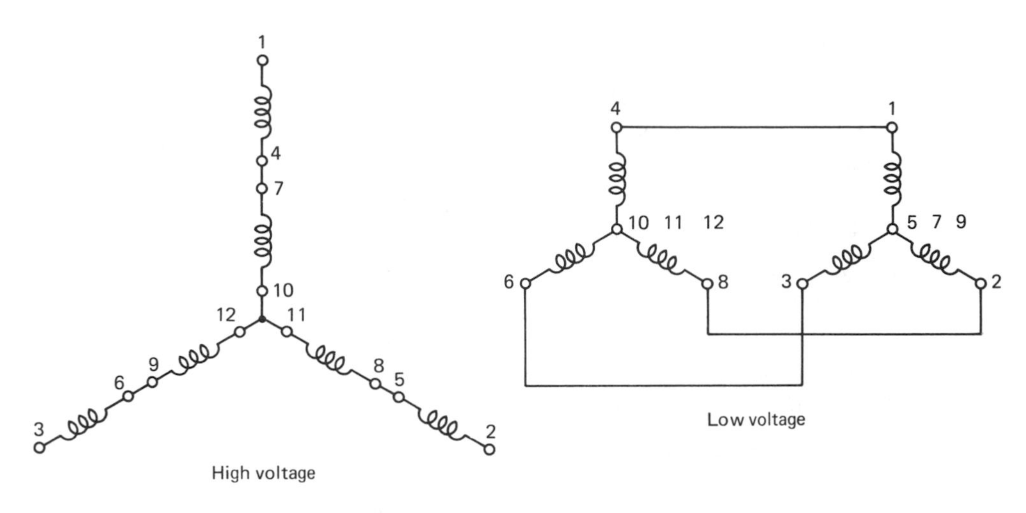

Figure 15-22 Twelve-lead motor connections.

test bench and change its load. This is easy in some cases but very expensive at other times. You must be the judge as to which test will be most beneficial to you.

Synchronous Motors

When a constant speed in a motor is essential, the synchronous motor is a good choice. The most popular example is an electric clock. The rotor of a synchronous is magnetized, and it locks itself in step with the rotating magnetic field of the motor. Since there is little or no slippage between the rotor and the magnetic field, the rotor speed matches the speed calculated by the motor characteristics. The large industrial version of the synchronous requires a separate DC source to magnetize the rotor.

SUMMARY

The study of motors goes beyond the principles of basic theory. Belt alignment, pulley size, and bearing wear are good examples of important factors that can affect the operation of a motor. Knowing the normal operation of any machine is essential to understanding and recognizing an abnormal operation or sequence when working with control circuits. This chapter was meant to help you acquire some of this basic knowledge. The next chapter on motor starters will examine the controls used to start and stop motors in various conditions.

EXERCISES

On a separate sheet of paper, complete a response for each question, statement, or problem listed below.

15.1. Define horsepower. How many electrical units is one horsepower equal to?

15.2. What is the relationship between the size and horsepower-rating of an electric motor?

15.3. Using mechanical terms, define horsepower.

15.4. List the two main categories of motors.

15.5. What is a main feature of the DC motor?

15.6. List and briefly describe the three configurations of a DC motor. What does each configuration determine?

15.7. How would you differentiate between a motor and a generator?

15.8. The voltage that is used to run the motor is called the ______.

15.9. What do the terms EMF and CEMF stand for?

16

Motor Starters

OBJECTIVES

Upon successful completion of this chapter, you should be able to

(1) describe the operation and construction of motor starters
(2) explain the uses of a starter
(3) discuss the procedure for selecting a motor starter
(4) describe the operation of three-phase and wound-rotor starters
(5) use an example to illustrate the applications for reduced-voltage starters and under-voltage control

INTRODUCTION

A primary device used to control motors is the motor starter. It not only provides a means for turning the motor on and off, but it can be used to regulate the currents and voltages that are applied to the motor.

MOTOR STARTER CONSTRUCTION

One of the best ways to understand how something works is to take it apart and look inside. Of course, you can not always do this, and it may not be practical to dismantle a perfectly good piece of equipment. If by chance you do replace a broken starter that cannot be repaired, then by all means disassemble the unit and examine how it works. There are some repairable parts in starters, and you will partially dismantle the unit to replace them. Take advantage of this time to get familiar with each type of starter that your firm is using. Replaceable items include contacts, overload heaters, and coils. To save you the problem of looking for a starter to examine, we will provide diagrams of the typical starter construction and explain its function. Three sections will be used to describe the operation and construction of a starter. These are the switching, coil, and overload sections.

Before we examine the parts of a starter, you should look at Figure 16-1. It shows the schematic diagram used to represent a motor that is connected to a line voltage with a starter. The dashed lines are not normally drawn on this type of schematic, but they are used here to show what sections of Figure 16-2 are in common.

Switching Section

Motor starters are switching elements between the line-voltage connections and the load connections to a motor. Figure 16-2 shows a pictorial diagram of each section of the starter. Once the line connections have been made, the switching section is the first part of the starter that current passes through. The electrical contacts of this section need to be rated for the load current drawn by the motor.

Figure 16-3 shows a common vertical arrangement of the switching contacts. Figure 16-3a represents the starter with the coil de-energized. It is important

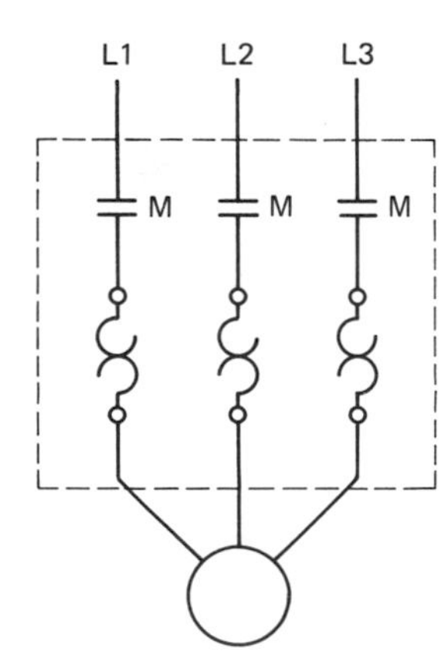

Figure 16-1 Motor starter schematic.

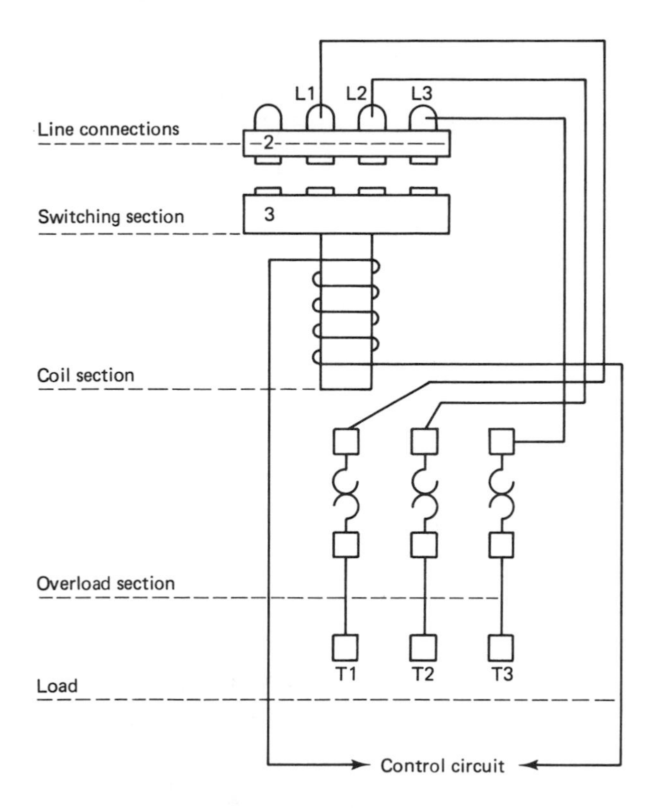

Figure 16-2 Pictorial diagram of starter.

to realize that the motor line current does not flow through the coil. This is an area of confusion for many people when they start learning about starters. If you treat them as two separate isolated circuits, you will avoid this problem. However, one feature is common between the two circuits. If you will examine Figure 16-2, you will see a set of electrical contacts that are labeled 2 and 3. These are the holding contacts we talked about in the earlier control diagrams. These contacts close when the coil is energized and are the same contacts that are labeled 2 and 3 in Figure 16-3. A second feature that is in common with both circuits is the overload contacts (they will be discussed later with the overload circuit).

Follow the current path on the diagram as the sequence of steps is given. We will start with Figure 16-3a.

1. Line voltage is applied to the line connection terminal of the starter. Only one of three possible phases are shown in each diagram.
2. The contact or switch is open at this time and no current flows.
3. Notice the contacts in the down position on the coil.
4. The control circuit is also in a de-energized position at this time.
   Now refer to Figure 16-3b.

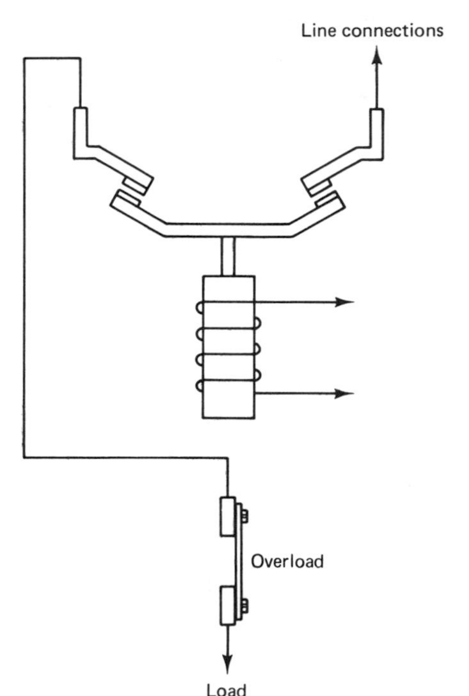

Figure 16-3 Vertical starter construction.

5. The start button is pushed and the control circuit is energized.
6. The coil has power applied to it; the resulting magnetic field pushes the plunger up along with the switching contacts.
7. Notice the new position of the contacts in relation to the coil.
8. Contacts 2 and 3 close at the same time and form a holding circuit.
9. The coil remains energized.
10. Current can now flow through the switching contacts to the top side of the overload assembly.
11. Current then flows through the overload assembly to the motor.

Figure 16-4 shows the same sequence of operations but with contacts mounted in a horizontal position. The control circuit is not shown in this diagram. You should repeat steps 1 to 11 for this circuit also.

Switching contacts that become pitted or corroded can sometimes be cleaned. The type of material used determines the cleaning method. Silver contacts should never be filed. Replacement with new contacts may be the best solution. Be sure to check the spring tension when they are replaced, otherwise you may be needing another set sooner than expected. Copper contacts can be cleaned with a file or

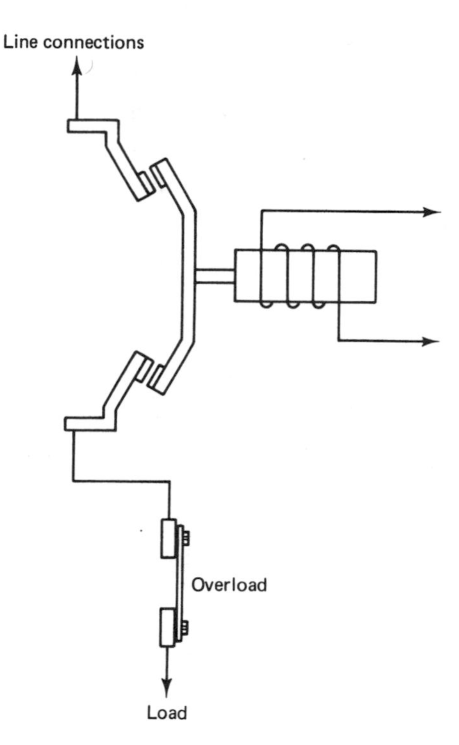

Figure 16-4 Horizontal starter construction.

fine sandpaper. Using a light film of isopropyl alcohol is still another alternative. Replacement contacts come with a set of springs. They should be used to replace the old springs, since spring tension can decrease with use.

Coil Section

The purpose of the coil section is to move the switching mechanism. It moves the plunger up or out to push the two sets of electrical contacts together. The contacts are spring loaded to oppose the force of the solenoid action of the coil. The plunger itself is spring loaded in the horizontal relay but can operate by gravity in the vertical relay, although it too is usually spring loaded to the OFF position. Spring loading the contacts to the OPEN position aids in their quick opening, which helps reduce arcing. Spring tension also assures a good solid connection between the two sets of contacts. Poor connections aid in carbon buildup and degrade the conductivity of the contacts. Coils are replaceable and can be changed when control circuit voltage requirements change. Coil voltages can be 24 VAC to 480 VAC and also come in DC current versions.

Another point of concern with coils is the current they use during startup. The inrush current has a surge of 6 to 10 times the normal operating current. This causes an excessive load on the control components. You must consider the rating of each pilot device when a coil is the load. The contacts will arc more than usual and eventually degrade the contact material coating. Carbon buildup will create a resistance and the heat will further degrade the performance of the contacts until they will eventually burn out. Industrial-grade pilot devices usually have sufficient ratings to prevent these problems, but you should be aware that using devices with insufficient ratings will cause you problems.

Overload Section

The overload section monitors the current supplied to the motor by the amount of heat created as it flows through it. When the magnitude of the current exceeds the preset rating of the heater unit, it mechanically opens the overload contacts in the control circuit. Figure 16-5 shows one set of overload contacts and their operation.

1. Current has a complete path through the switching part of the starter to the interchangeable heater units.

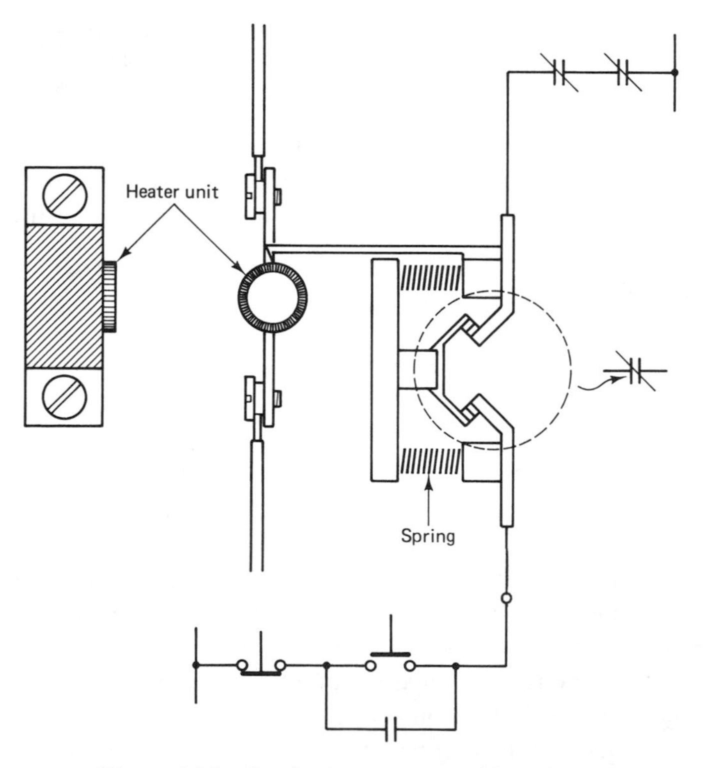

Figure 16-5 Overload contact assembly—closed.

2. The heater units have small heat coils built into them.
3. A wheel mounted on the side of the heater unit has its shaft embedded into a solder pot. The normally cool temperatures of normal current are not sufficient to melt this solder, and the wheel is not free to turn.
4. A mechanical latch is engaged into part of the wheel and holds a spring-loaded assembly that keeps the overload contacts in a closed position.
5. When motor current exceeds heater unit ratings, the solder begins to melt and the mechanical latch is no longer able to hold the spring-loaded overload contacts in a CLOSED position.
6. The wheel spins and the contacts open (See Figure 16-6).
7. When the overload contacts open, the control circuit is broken, and the starter coil is no longer energized.
8. The coil no longer has a magnetic field strong enough to close the switching section, and it drops to the de-energized position, opening the contacts.
9. This turns the motor OFF.

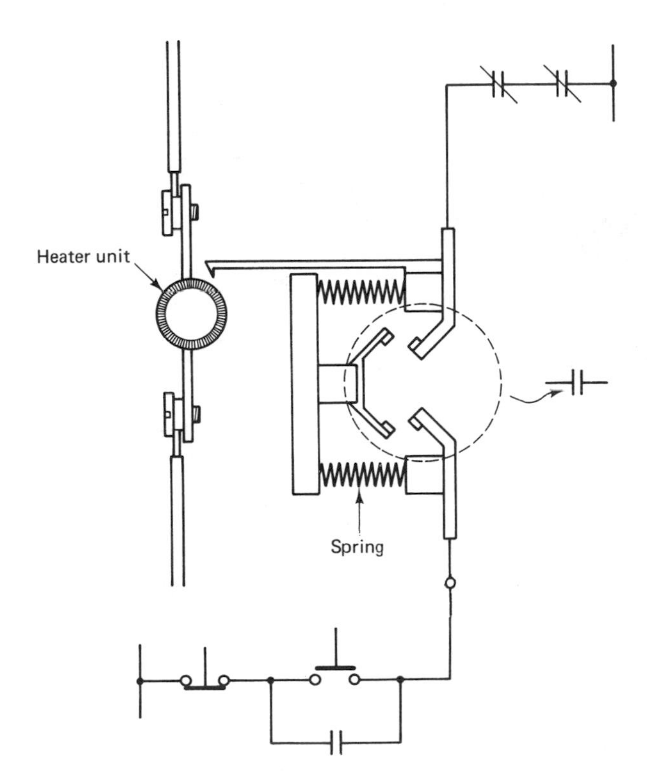

Figure 16-6 Overload contact assembly—open.

10. After the motor has been off, the heater unit cools and the solder pot again solidifies and the wheel is no longer free wheeling.
11. The overload contact mechanism can be reset by pushing a manual switch.

USES FOR STARTERS

Starters are relays with overload protection added. They can serve several purposes. Besides the switching function, they do provide a protection factor for the motor by preventing it from developing excessive heat during an overload condition. They also allow changing speeds and direction.

Forward-Reverse Starter

Figure 16-7 illustrates two starters combined into a single unit for changing direction in a three-phase motor. One method of reversing a three-phase motor is to interchange any two of the power leads that supply current to the motor. To do this manually every time you wish to change rotation would be very bothersome. The forward-reverse starter does this mechanically. Follow through on Figure 16-7 as we describe the sequence of operation for the starter. You may also wish to review the control diagram for this starter on Figure 3-11.

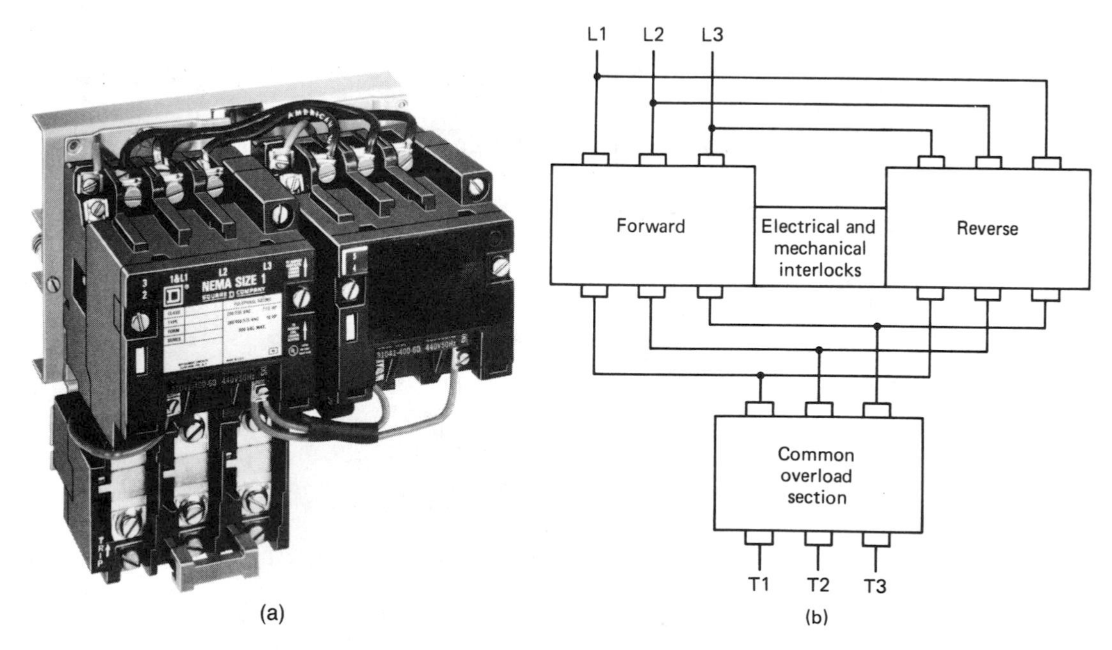

Figure 16-7 (a) Forward reverse starter (courtesy of Square D Company). (b) Forward and reverse starter.

1. Three-phase power is supplied by line connections on one side of the switching section of a starter.
2. The forward button is pushed, and the control circuit energizes the forward coil, closing the forward switching section.
3. Current flows through the starter to the top side of the overload section, through the overload heaters and to the motor.
4. Only one overload assembly is required, since both forward and reverse starters cannot be energized at the same time. This is because of their interlock circuit, which may be electrical or mechanical or both. The same amount of current flows when the motor is running in either direction.
5. Again, the interlock circuit prevents the opposite starter from being energized once the other starter is engaged.
6. If the forward unit is off, the reverse starter circuit can be energized.
7. Notice the arrangement of the wires on the top of the two starters.
8. The forward starter has the power connections going from L1, L2, and L3.
9. The reverse starter has the power connections going from L3, L2, and L1. This satisfies the requirement of changing any two leads to change rotation.
10. The interlock circuit also prevents back-feeding power through the other starter unit, which would short circuit the power connections.

PUMP ALTERNATING CIRCUIT

This circuit is used to alternate two pumps in order that they both get the same amount of usage over their lifetime. Although it involves two starters, the key to its circuit operation is the alternator unit. The alternator in Figure 16-8 is composed of a relay that closes contacts A1 on one cycle and alternately contacts A2 on the next cycle. Follow the circuit through as we describe its operation.

1. Float switch F1 closes as water enters a tank. This supplies power to the circuit.
2. As the water level continues to rise, float switch F2 closes and energizes the alternator coil A.
3. Assuming the contacts are in the position as shown after the coil A is energized, current can flow through contact A1 to coil P1, which energizes pump number one. Contact P1 closes and forms a holding circuit for pump one.
4. If the water level continues to rise and pump one cannot keep up, then float switch F3 will close and start pump two. Contact P2 closes, forming a holding contact for pump two.

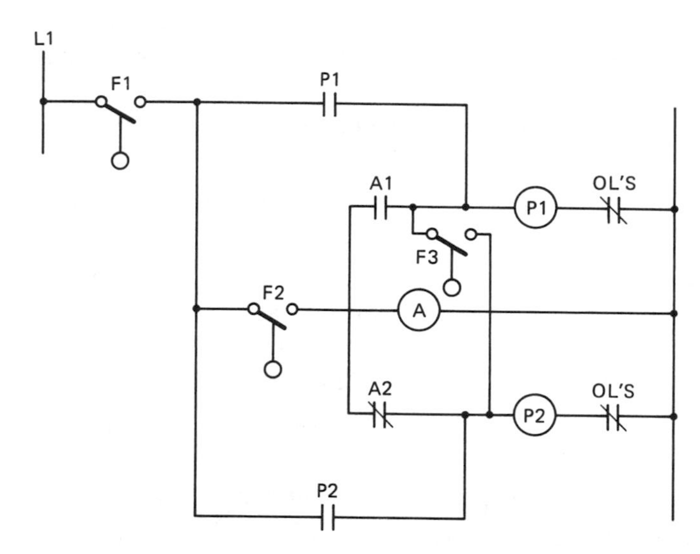

Figure 16-8 Pump alternating circuit.

5. Now, as the water level decreases, float switch F3 opens but has no effect due to the holding circuits.
6. If the water level continues to fall, float switch F2 will open, but it too has no effect due to the holding circuits.
7. Finally, float switch F1 opens and the entire circuit is de-energized—the pumps are turned off.
8. This completes one cycle.
9. The next time float switch F1 closes, the alternator switches and contacts A2 are now closed and contacts A1 remain open.
10. Pump 2 will start first and pump 1 will follow.
11. If float switch F3 does not close, then only one pump will run.

FACTORS FOR STARTER SELECTION

Starter Selection

Selection of a starter begins with reading the nameplate of the motor you are going to protect. Nameplate data is the basis for choosing the appropriate starter because the most accurate source of motor characteristics is directly from the manufacturer. These may include horsepower, full-load current, temperature rating, duty cycle, speed, voltage ratings, enclosure type, and service factor. Two of these items are generally used to select starters from tables provided by firms that manufacture the units. Horsepower and voltage ratings when considered together can be grouped into standard NEMA (National Electrical Manufacturing Association) categories. They range from size 00 to 9. An alternate method is to use the continuous-current rating of the motor. Each NEMA size also corresponds to a maximum current rating. This will allow you to choose an appropriate starter for continuous-duty operations. Continuous duty is operating at a stabilized condition for an extended period of time (running for more than three hours). If the duty cycle is something other than continuous duty, this usually requires using a larger unit. The manufacturer's bulletin can provide you with the correct item for your particular use. Coil selection is based on the control voltage used.

Overload Protection

Without overload protection a starter would just be a large relay. This is the whole purpose of buying a starter. Motors are expensive to buy and repair, and this does not include the cost of down time in an assembly system. Four items are of importance for calculating the correct overload heater unit. Rated current, operating temperature, service factor, and ambient temperature determine if the size of the heater unit selected will actually protect your motor from an overload condition. An overload condition does not include short-circuit conditions, but is considered to be the operation at an excessive current condition for a long enough period of time that it will damage the motor.

Heater units are selected from the manufacturer's tables and must be designed for the particular starter unit you are using. Using the full-load current, it is a simple matter of matching the current with the heater unit number. This assumes several things. First, the motor is operating at a maximum allowable temperature rise of 40 degrees Centigrade and a service factor of 1.15. Many times the starter unit is not located near the motor. If this is the case and the temperature at the starter is higher than that of the motor, then a next higher rating is selected; if it is lower, then the next lower size rating should be used. We will give an example of two methods of calculating the current level at which an overload unit should operate.

1. Determine the recommended heater unit based on full-load current of the motor.
2. The product of the minimum current rating of the selected heater unit times 1.25 will indicate the operating current.
3. Ambient temperature correction is 1.00 times this rating if the temperature is forty degrees Centigrade.
4. If the temperature is other than forty degrees Centigrade use the correction factor determined from Figure 16-9.

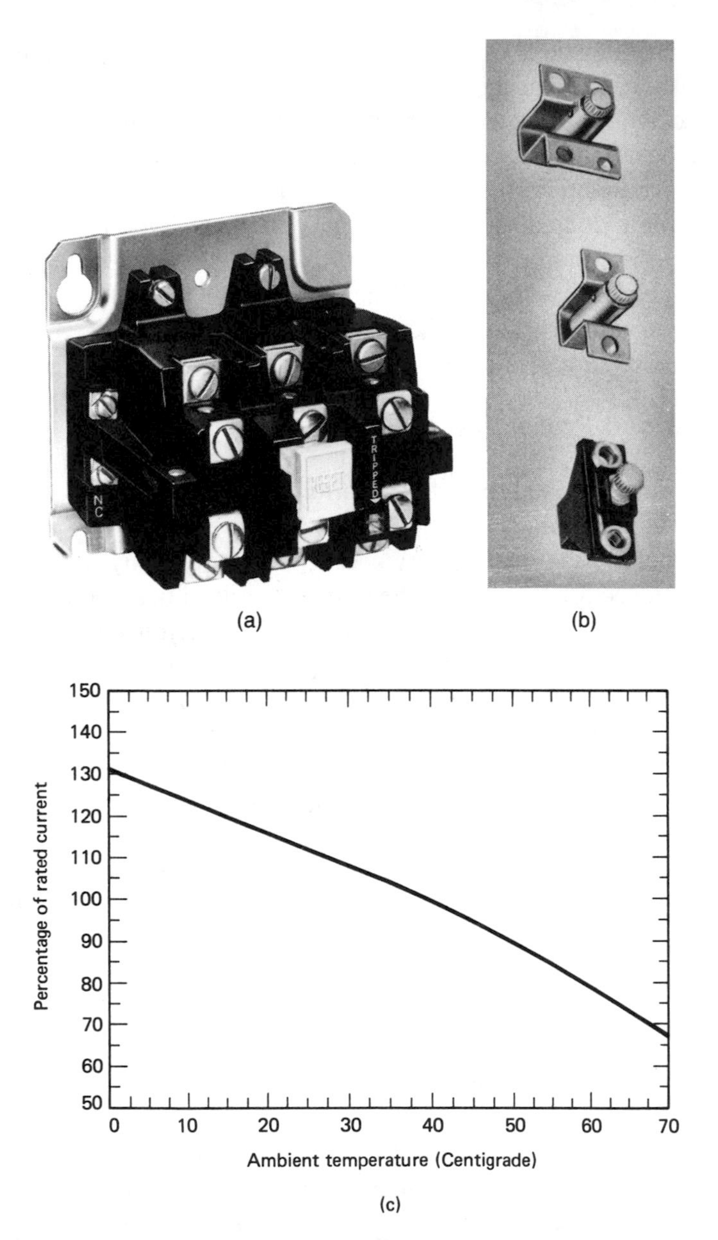

Figure 16-9 (a) Overload assembly without overload units. (b) Typical overload units. (c) Heater unit correction curve. (Photos courtesy of Allen-Bradley).

Let's assume a full load current of 20 A and a heater unit with a working range of 18.0 to 22.0 A. In this example if the ambient temperature was 52 degrees Centigrade, the correction factor would be 0.9. The operating current for the overload unit would then be 20.25 A (18.0 A × 1.25 × 0.9).

Single-Phase Operation

Single-phase operation with a starter can be accomplished using either a single-phase starter or a three-phase starter. Single-phase units have two sets of contacts. One line of a single-phase circuit must have overload protection. Selection of these units is the same as described in the selection process.

Three-Phase Operation

Many of the circuits described throughout the book have been three phase. The process of selecting a starter is not unique except that using three overload units instead of two provides added protection for the motor. Unbalanced currents in three-phase circuits can cause excessive currents in the remaining windings. Single-phasing can degrade the insulation of the motor windings by excessive heat. Eventually the insulation can break down and the motor can short-circuit. No additional selection procedures are necessary beyond the topics in this chapter.

Wound-Rotor Starters

The wound-rotor motor is a squirrel-cage motor with a series of resistances added to control the speed and torque developed during startup. Figure 16-10 shows a simple diagram of the wound-rotor motor. A series of relays that short-circuit the resistances for each speed can be used to step changes in speed. A variable speed control may also be used. The entire control circuit has been eliminated from the diagram, since we have discussed sequence starts and interlocks in previous chapters. Some interlocks are required to assure that the motor starts with all resistance in the circuit. Examine Figure 16-10 as we describe the operation of this starting circuit.

1. The first step is to energize starter M and apply power to the circuit.
2. At this time all resistance is in the circuit and the motor runs at slow speed.
3. When speed two is selected, part of the resistance is shorted out of the circuit and the motor speeds up. The speed selection may be done manually or with a timing circuit.

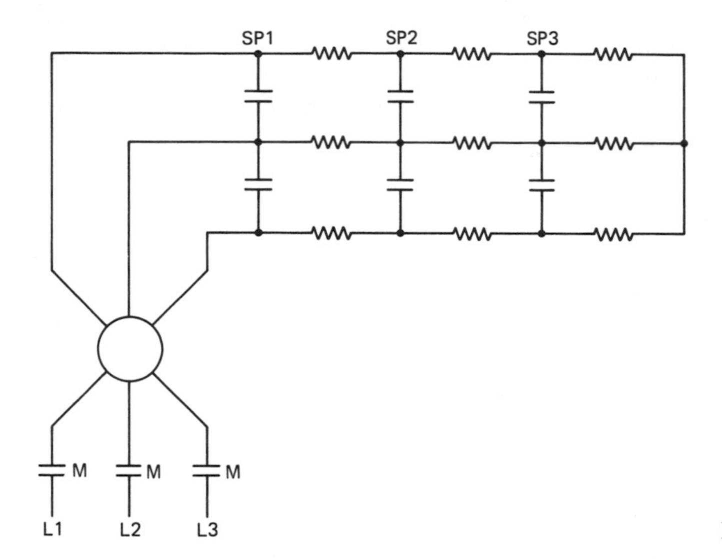

Figure 16-10 Wound-rotor motor.

4. Speed three further reduces the resistance and the speed increases.
5. At speed four all resistance is effectively out of the circuit, and the motor runs as a squirrel-cage motor.

Reduced-Voltage Starters

When large motors in industry are started, the initial large surge of current not only goes through the motor but through the power lines supplying the factory. This surge is experienced all the way to the power plant. To maintain voltage and current levels at a stable level requires the power company to limit surges on their lines. One method of doing this is to require that owners of large motors to use reduced-voltage starters. By doing this the initial surge of the motor is limited, and power line currents remain at a relatively stable level.

Figure 16-11 shows a typical reduced-voltage starter. Control of the motor can either be manual or automatic. Manual control is less expensive, and since the number of times these large motors are stopped and started on a given day is limited, manual control is acceptable. Follow the circuit's operation as we describe the steps in sequence.

1. Starter M supplied three-phase power to the motor.
2. Current must flow through the resistors in each phase as the motor begins to start; this limits the current flowing to the motor.

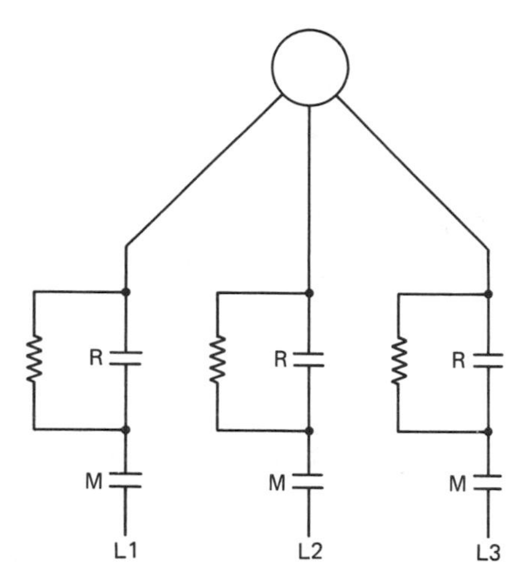

Figure 16-11 Reduced-voltage starter.

3. After a given time, the run contactor is energized. This time is based either on a tachometer giving an indication of the motor's being at full speed for this condition or, more than likely, someone with a good ear listening to the motor speed. This may sound inaccurate, but a large motor has a definite sound as it accelerates and then settles at a given speed.
4. Switching the run relay in the circuit removes all resistance, and the motor is allowed to accelerate to its normal running speed.
5. The current required to accelerate from an intermediate speed to full speed is considerably different than from a full stop condition to full speed. This is the basis for the reduced-voltage starter.

Undervoltage Control

Another important factor in the control of motors is protection against undervoltage conditions. The circuit in Figure 16-12 shows one method of providing this protection.

1. When the start button is pushed, current flows through the undervoltage relay, UV, and the diodes.
2. Relay UV picks up and coil M is allowed to energize, starting the motor.
3. While the motor is running, the capacitor is being charged to a sufficient level to maintain coil UV energized.
4. Should the voltage level drop, coil M will drop out.

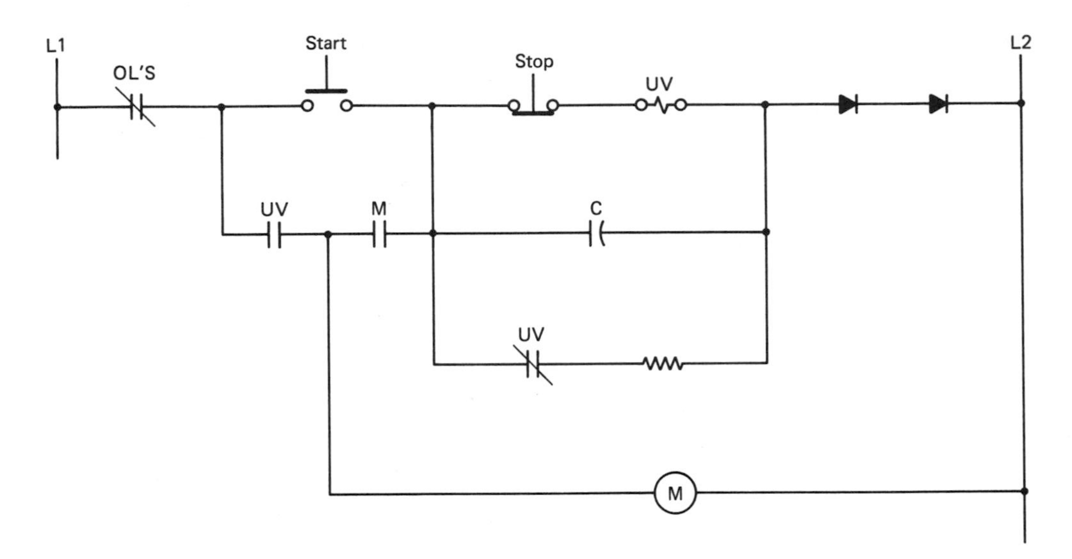

Figure 16-12 Undervoltage protection circuit.

5. This leaves a small circuit within the control circuit consisting of a capacitor, branch two with the normally closed contacts of UV (which is open because the coil is still energized), a resistor in series, and the stop button with the coil of relay UV in series. All three are in parallel.
6. When coil M dropped out, contact M opened and isolated this circuit.
7. This circuit allows the capacitor to discharge through the relay coil.
8. The value of this capacitor forms a timing circuit which will keep coil UV energized until the voltage level decays to a magnitude insufficient to keep coil UV energized.
9. Should the voltage level increase before coil UV de-energizes, coil M will be allowed to energize and contact M will pick up.
10. This keeps the circuit energized during short undervoltage transients and allows the motor to de-energize during these transients.
11. A long-term undervoltage condition will shut down the entire circuit.

SUMMARY

Several important features of motor starters have been provided in this chapter. Many more maintenance factors should also be undertaken. These include periodic testing of motor insulation, bearing changes, and routine cleaning of motor starter cabinets and motor housings. Proper care of your equipment will not only prolong the life of the units but will also help eliminate potential problems when they are not only time consuming to repair but costly. Our next chapter investigates some factors that are important when selecting relays as control elements.

EXERCISES

On a separate sheet of paper, complete a response for each question, statement, or problem listed below.

16.1. What is a motor starter?

16.2. List the three sections of a motor starter.

16.3. What is the purpose of the coil section in a motor starter?

16.4. List two primary uses for a starter.

16.5. What is the purpose of the pump alternating circuit?

16.6. Describe primary considerations for selecting a starter.

16.7. Without an overload protection, what purpose would a starter serve?

16.8. How would you use a starter for single phase operation?

16.9. Compared to a single-phase starter, what is the advantage of using a three-phase starter?

16.10. Another name for a squirrel cage motor is ______.

16.11. What is the function of the reduced voltage starter?

16.12. How would you protect against undervoltage conditions?

17

Relay Selection

OBJECTIVES

Upon successful completion of this chapter, you should be able to

(1) explain the terms *reliability* and *life expectancy* with respect to relays
(2) list components of a relay and discuss how relays are constructed
(3) explain typical relay problems
(4) describe *drop-out* or *pull-in* currents
(5) describe solid-state relays and explain how they are different from, or similar to, mechanical relays

INTRODUCTION

There is little doubt that one of the most common control elements is the relay. No other single element can provide the simplicity and yet the versatility of a relay.

You have already been exposed to the many different mechanisms used to actuate relays. This chapter will examine in a little closer detail relay construction and characteristics that many times are overlooked when using them. Hopefully you will gain a better appreciation for the proper selection of a relay and make the characteristics it has work for you. We will deal with mechanical relays, since there are so many in use, and they are also the most misused.

RELAY CONSTRUCTION

If you will recall from our discussion on starters, the starter can be considered a relay with overload protection. One section of it was the switching section. In a relay, the same basic functions are still present—the switching section and the coil section. A common construction technique is used on many relays, and it is shown in Figure 17-1. Although there are many different variations of relays, this diagram displays the underlying principles of all relays. One feature that will be important to keep in mind any time you are working with a relay diagram is that all contact positions are shown in the de-energized position unless otherwise noted. This serves as a common point of reference for all users and designers of relay products.

The major parts of a relay are the coil, the lever mechanism, and the relay contacts. Each part is engineered for a particular reason. It is not just put together because it works, but specific purposes for using a relay may require each separate component to have a unique design to meet a set of design specifications. Refer to Figure 17-1 as we examine the normal operation of a typical relay.

1. AC or DC power is supplied to a coil which creates a magnetic field in the iron core.
2. The lever mechanism is held in a normal position with spring tension.
3. When the magnctic field of the coil is energized, it overcomes this spring tension and pulls the lever toward the coil.
4. This downward movement opens the set of normally closed contacts and closes the normally open contacts.
5. When power is removed from the coil, spring tension pulls the lever back, and the contacts return to their de-energized position.

This simple operation we have just described conceals a great many engineering principles. We will try to simplify this for you as much as possible as we examine each component.

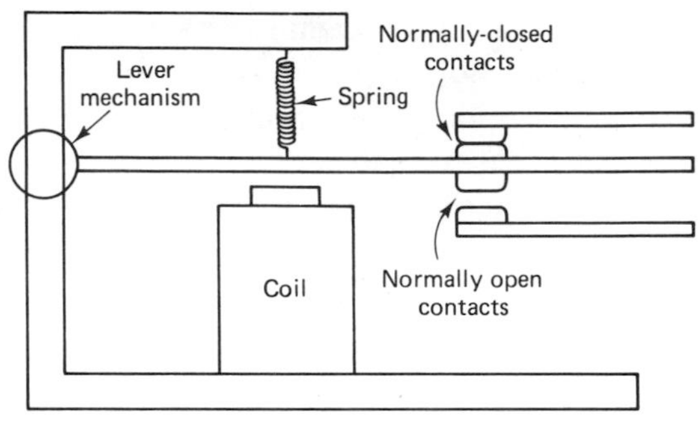

(a) De-energized

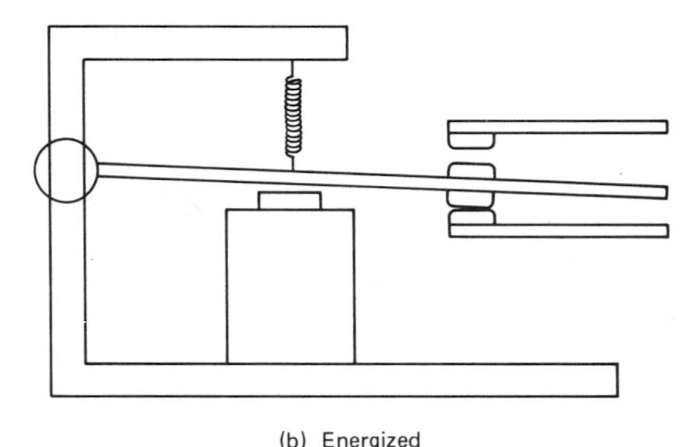

(b) Energized

Figure 17-1 Basic relay construction.

Relay Coil

We will begin examining the relay by starting with the coil. The coil is the fundamental unit required to develop a force strong enough to move the relay mechanism. Figure 17-2a shows how current flowing through a wire that has many loops can be used to concentrate a magnetic field. To increase the efficiency of this coil, an iron core is added, and the magnetic field will concentrate its flux in this core. The magnetic field also creates polarities on the ends of this core, as shown in Figure 17-2b. We now have an electromagnet. DC-operated coils have few problems with maintaining a constant field strength. AC coils have a problem when the field reverses and the polarity of the electromagnet also reverses. To overcome this, a shading ring is used to help retain the residual magnetism of the first half cycle of current long enough during the second half cycle of current to maintain an electromagnetic force in the same direction as a DC coil.

The strength of a coil is expressed in terms of ampere-turns. We will assume that our coil has constant values for the iron core, spacing of the windings, and physical dimensions, all of which can affect our basic calculations. Let's use this

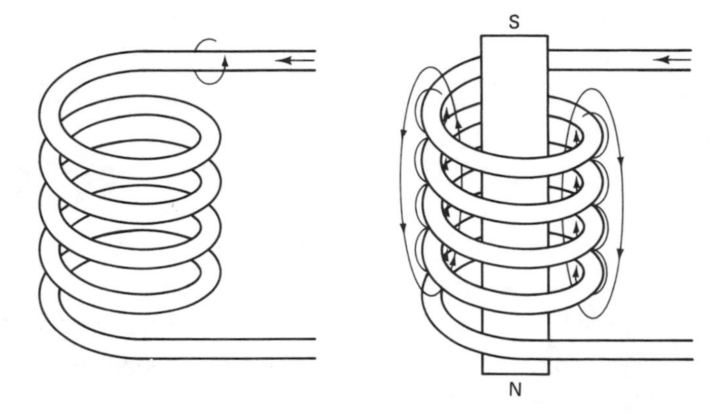

Figure 17-2 Magnetic field of a coil.

formula: coil strength (ampere-turns) = $N \times I$, where N is the number of turns, and I is the amount of current flow (in amperes). There is a direct relationship between coil strength and the number of turns or the amperage in the coil. Figure 17-3 illustrates how different currents and number of turns affect the ampere-turn values. Increasing the number of turns is one method of allowing an increase in voltage and maintaining a fixed current for a given coil, but this also allows the force to increase. This is not usually necessary, since the mechanical forces required to operate the relay are not relevant to control circuit voltages. Notice that letting the current increase by reducing the number of turns will allow the ampere-turn rating to remain constant. The strength of the magnetic field required will determine which factor will be adjusted during the design of coil when voltages are changed.

Figure 17-4a represents a coil with fixed characteristics. Assuming negligi-

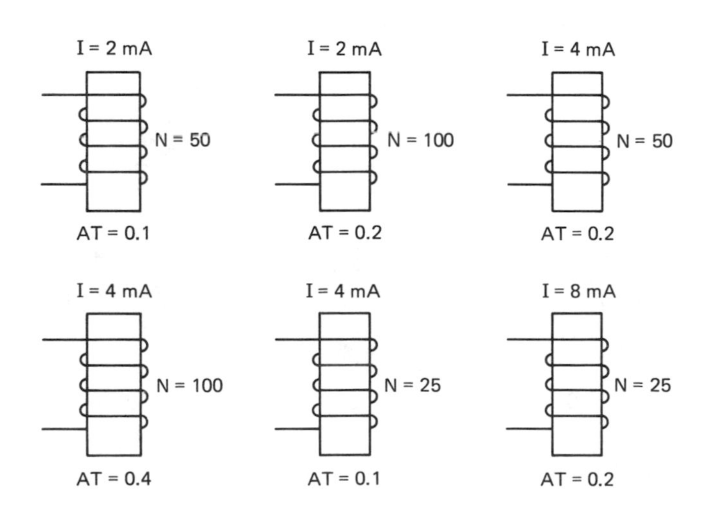

Figure 17-3 Ampere-turn calculations.

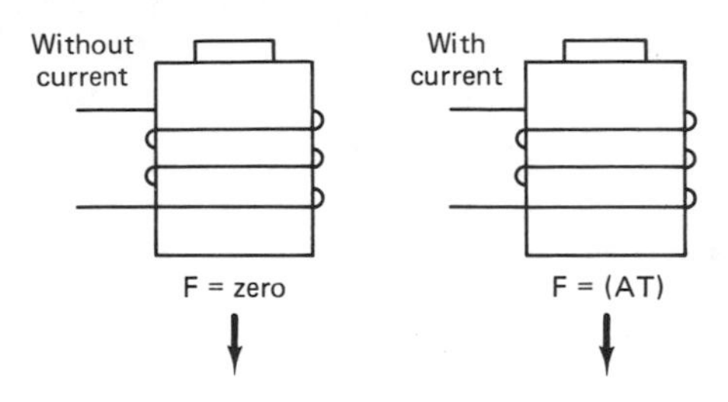

Figure 17-4 Coil forces.

ble residual magnetism, we will assign a magnitude of zero to the force created when no current flows in the coil. Figure 17-4b shows the same coil energized. The force created is shown with the direction indicated.

Figure 17-5a illustrates a theoretical arrangement of how the forces could be aligned in a relay. Force one, F1, represents the spring tension. This force should have magnitude of sufficient strength to counteract the weight of the lever plus the additional force required to assure good contact between the normally closed contacts. Force two, F2, also helps keep the contacts closed, which helps keep them from arcing. Force three, F3, is the force created by the coil when it is energized. This force must be of sufficient strength to overcome spring tension and prevent contact bounce when the normally open contacts close. It must also assure good seal in connections between the contacts.

Figure 17-5b summarizes all the forces involved when the relay is de-energized. F1 and F2 are the only factors that need be considered. In Figure 17-5c, F1 remains as a constant. Force three must therefore overcome F1 and F4 to operate. Force two is not a factor when the coil is energized. There is very little that can be done to control the forces, since they are part of the design process. However, several things do happen that can affect the operation of the relay. Consider what effect low voltage on a coil would have on the ampere-turn value. Decreasing the force of F3 can prevent the coil from being fully energized, limit the contact force between the normally open contacts, and increase the effect of contact bounce. A high-voltage condition will eventually cause the coil either to open or short-circuit when the insulation on the wire breaks down. Hopefully you can see it is important to maintain the normal operating voltage of the relay for its best performance.

Figure 17-6 shows a simple solution that can help counteract part of the low voltage problem. Moving the coil farther out in the lever reduces the required force to move it when under the same conditions. Again this is a design feature that we cannot control.

One final point is the possibility of someone's installing the wrong coil in a starter. If the ampere-turn value is too small for the unit, we will have the same symptoms as an undervoltage condition on a relay.

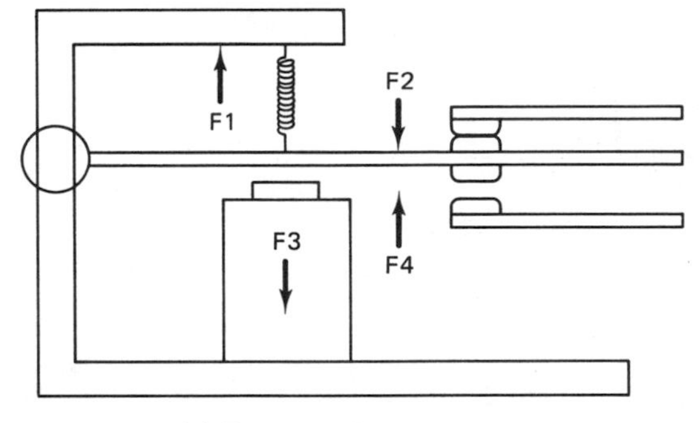

(a) Theoretical force alignment

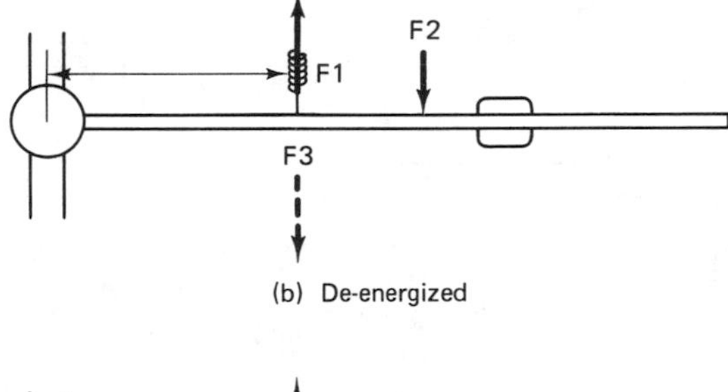

(b) De-energized

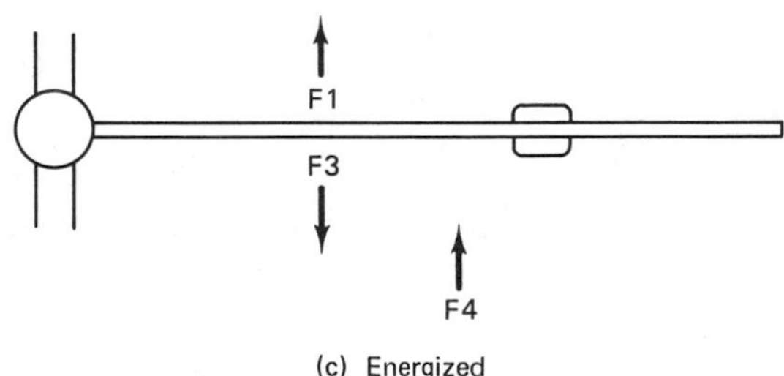

(c) Energized

Figure 17-5 Relay forces during operation.

Contacts

The type of contacts used also is very important. Some relay manufacturers offer a choice of contacts for some of their relay models. Some metals are more conductive due to their physical characteristics. Such metals used as contacts are silver, silver cadmium oxide, palladium, platinum, and gold. Each has its own unique property that makes it more suitable than others for given situations. Silver is popular more for the economic aspect than any other reason. In addition, it does have good electrical characteristics and is easily shaped. Arcing is a problem for contacts at high currents. Silver cadmium oxide coatings resist some of these problems because of their ability not to oxidize readily at high current levels. Palladium and platinum are good for highly corrosive environments but have

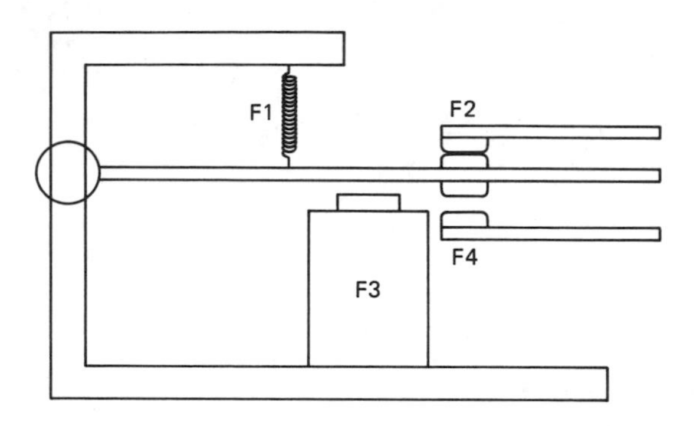

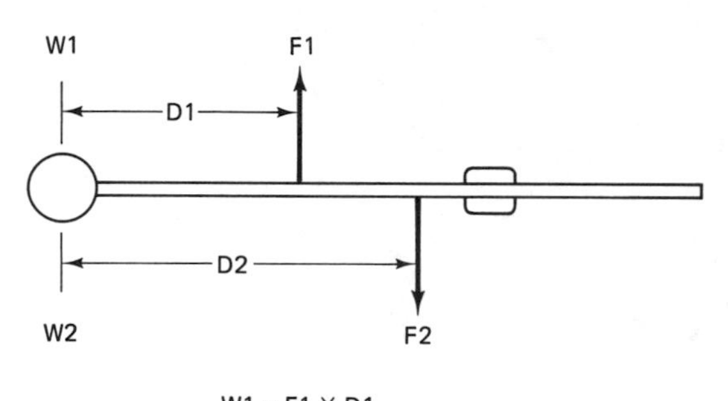

W1 = F1 X D1
W2 = F2 X D2
W2 must be greater than W1

Figure 17-6 Improved design of relay.

limited current-handling capabilities. Gold is another low-amperage rated contact material. Its high cost limits its use only to specialized circuits that need the reliability and are worth the added cost.

Operating voltages and arcing go together when contacts are considered. Excessive arcing can melt the contact material and enhance the buildup of oxides which create resistance and heat. Several methods of reducing contact arc include not only choosing the correct contact material but contact arrangement, spacing, and voltage considerations. It can be very beneficial for you to consult with your relay supplier when you have problems and special applications for relay products. Most firms have engineering consultants that are more than willing to provide application assistance for their products.

Contact bounce can be a problem with some relays, especially when they control inputs to computers. Bouncing contacts can represent multiple inputs to a computer. Low-bounce relays or solid-state relays should be considered in these circuits.

Contact life should also be considered during the selection process. Factors that can shorten the useful life of contacts are susceptibility to corrosion, type of load, type of voltage, and duty cycle—to mention a few. Any one item will shorten the life factor, but they all can have a cumulative effect when combined.

Relay Form

Numerous contact arrangements are possible with relays. Some of the combinations are similar to those of toggle switches that were discussed in Chapter 1. Figure 17-7 shows relay versions of the SPST and SPDT switches. The contact arrangements have been standardized by the United States of America Standards Institute (USASI). Those given in Figure 17-7 show the *Form Type* designation given to them by the USASI. There are several other form types designated, but only the more commonly used symbols are shown here. If you consult with a firm about relay applications, it would be beneficial to be familiar with form designations. It will provide you both with a common reference; this is especially true with more complex types of relay contact arrangements.

Terminal Connections

A variety of connection methods are available for relays. Screw-terminal connections are popular with the higher-current relays. Solderless-clamping lugs are popular since they do not require a terminal lug to be installed and can accommodate more than one wire. Smaller wires are sometimes soldered to the terminals. Figure 17-8 illustrates these three methods. The popularity of plug-in type relays has resulted from their ease in replacement. Figure 17-9 shows the base of a plug-in relay and its wiring diagram. Since they plug into a socket unit, the wires are

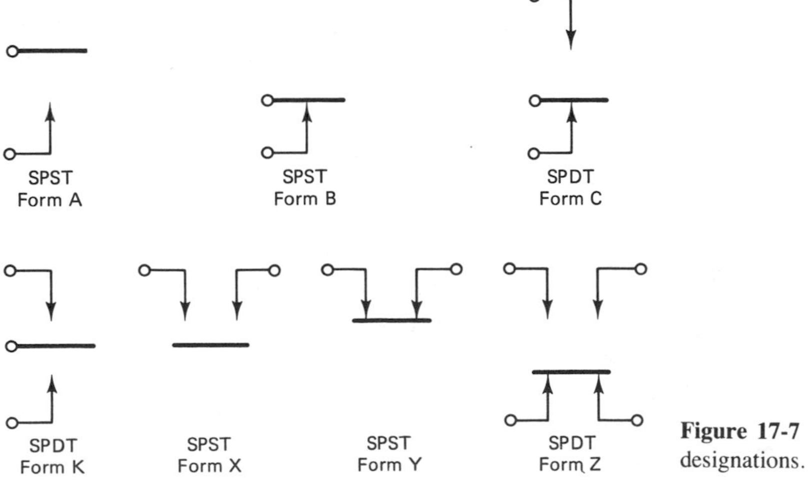

Figure 17-7 Relay form designations.

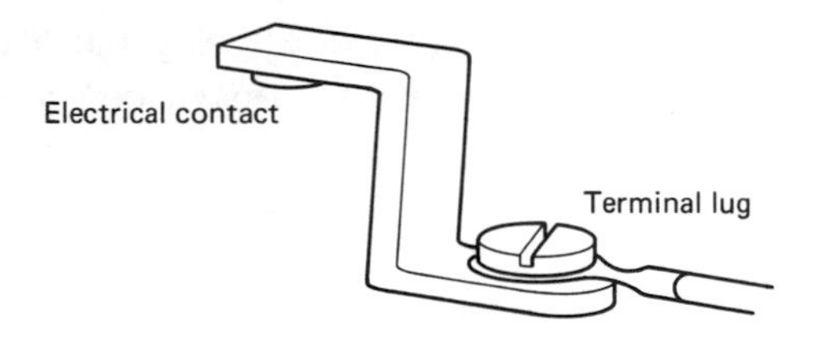

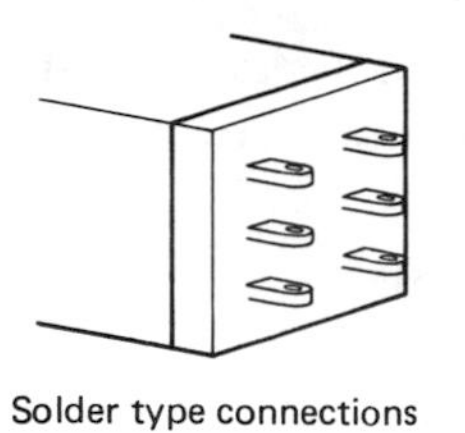

Figure 17-8 Various terminal connections.

not connected directly to the relay itself. Using the wiring diagram, wires are connected to the base unit by terminal numbers. Sometimes it is necessary to check continuity of a plug-in relay while it is out of the socket. This can be done by using the wiring diagram and knowing the terminal arrangement on the relay. Terminals on the relay are numbered from 1 to 8 or 11, depending on the number of contacts. Numbering begins at a key and goes clockwise when viewed from the bottom of the relay. Remember the relay is de-energized when checking continuity, and it does not indicate the condition of the relay once it is energized. If in doubt it is a simple matter to just replace the relay and then check the circuit for proper operation.

Life Expectancy

When a relay is installed, a certain degree of trouble-free operation is expected. Assuming a properly engineered product, the variables that can degrade operational life depend on the user's following these design specifications. Using a relay out of specification guidelines is one of the prime reasons for relay failure. To help you stay within these limits, you should purchase a relay using a set of specifications of your own. These specifications should include what voltages are required for both the contacts and the coil, what current is passing through the contacts on closing (inrush), and when the load is on the line (steady state). The

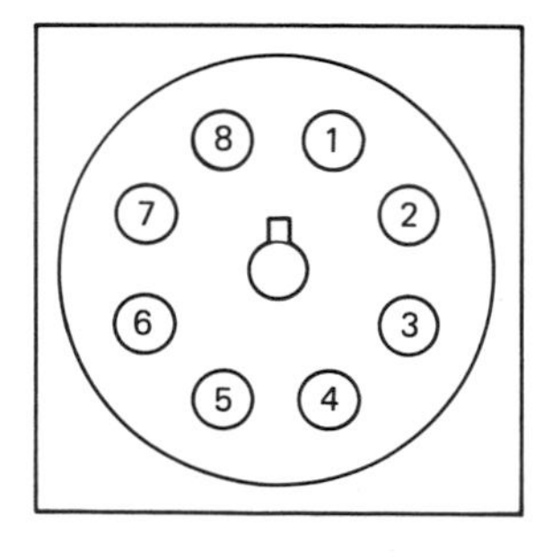

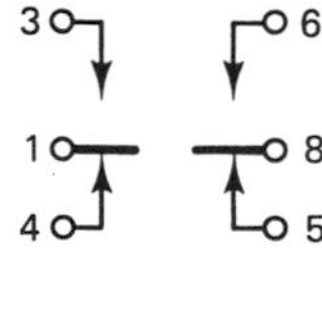

Figure 17-9 Relay socket and pin diagram.

duty cycle, or frequency of operation, can shorten the life of a relay. It may be necessary to use a higher-rated relay if the frequency of operations is high. Environmental factors may make it necessary to enclose the relay to protect the contacts from corrosion. Consider each of these factors as you choose a relay, and you can expect a greater useful lifetime out of your equipment.

Reliability

Several terms are used to describe the reliability of a relay. One of the terms is *Mean-Cycles-Between-Failures* (MCBF). This gives an indication of the number of operations that can be expected before a failure occurs. This number should be a very large number if the reliability is to be high. A small number would indicate poor reliability. Mathematically, this is expressed as MCBF = (Number of Attempts)/(Number of Failures). One thousand attempts with one failure would be 1,000 cycles between failures. One thousand attempts with four failures would be 250 cycles between failures. The greater the attempts and the fewer failures that occur will improve the MCBF factor. Standard design features have made relays very dependable items. When it is critical that the failure rate be reduced beyond the normal manufacturing specifications, some means must be found to increase the reliability. Relay designers will have to add materials and equipment to their original design that improves the reliability of each individual component. The total reliability of any production item is a function of all the individual reliabilities. No doubt you have heard of one bad apple spoiling the whole bunch. This same principle holds true here also. Improved reliability does not come without its price. Quality items cost more to produce, and non-standard products that must be specially made will include the additional cost of designing them.

One method of improving the reliability is to have multiple items doing the same operation at the same time. Even though each individual item may have a low reliability, the effect of all the items together increases the reliability factor. This is also called redundancy. Normal operations do not require this added expense, however, since the added expense of buying a better-quality relay will generally outweigh the price of buying many separate relays. The bottom line is what the budget will allow and what the price of failure is to you. Purchasing a twenty-dollar relay that will cost you two thousand dollars an hour every six months is not as wise as buying a hundred-dollar relay that will fail only once every two years. Buying only high-quality equipment will reduce the number of failures in a system. For the same reason, using poor-quality equipment would also increase the number of failures. There is a trade off that must be made between cost and reliability, but only you can decide where this point is in your process.

PROBLEMS WITH RELAYS

Misapplication is the greatest source of problems with relays. Choosing the wrong relay for a given purpose is almost a guarantee of failure. High-current-rated contacts have an arc that acts to clean some of the film built up on them during operation. Without this arc the contact resistance will increase. This can occur when small current values are used on these relays.

Nonresistive loads can create surges of current many times the normal steady state values. Relays that must switch these types of loads should be capable of handling surge currents as much as six times the normal steady-state values. In addition, current-limiting devices, such as varistors, may be required to reduce contact wear. Figure 17-10 shows some typical arc-suppression circuits used in AC and DC circuits.

Cleaning Relays

Relays are cleaned in the same manner as starters. Not all relay contacts are replaceable, however. Many are enclosed in protective plastic or metal cases to protect them from environmental factors. This still does not protect them from misuse. Contacts in these types of enclosures are repaired by replacing the entire unit. Properly applied use of relay systems will allow many trouble-free operations.

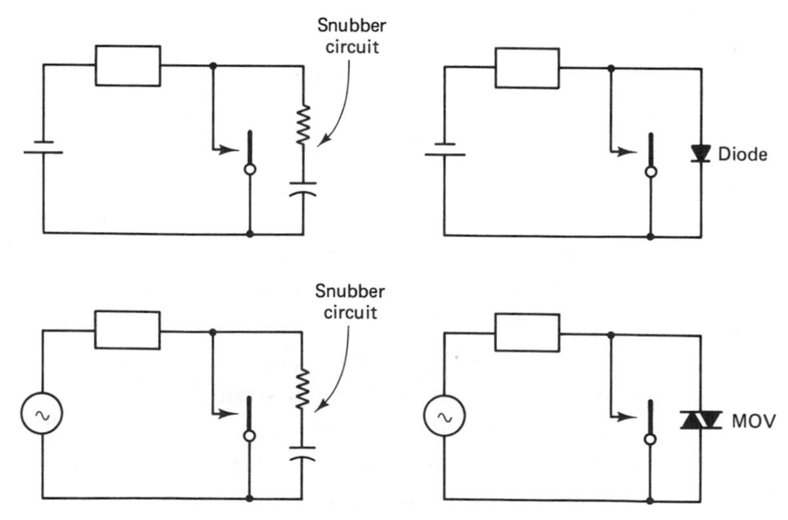

Figure 17-10 Arc suppression circuits.

Relay Terms

There are a few terms used to describe operation of relays that are important to understand. These are pull-in current and drop-out current. *Pull-in* current refers to the amount of current required to have the relay fully energized. This voltage must overcome all the forces that keep the relay normally open. Once the relay is closed, the operating current is the required current to keep it in an energized state. If the current is decreased, there is a value at which the relay will no longer remain energized. This is the *drop-out* current.

Solid-State Relays

Solid-state relays are electronic switching devices that may or may not use any moving parts. If there are moving parts, they are called *hybrid* relays. The advantage of this type is that even though there are moving parts in the circuit they are sealed within the unit and are not subject to corrosion. They may also be made of smaller and more reliable products, since they are not usually required to switch the full current of the load but instead only the control current. A pure solid-state relay will have no moving parts. All components are solid-state devices like transistors, silicon-controlled rectifiers or triacs. These are also found in part of the hybrid circuit as well. Figure 17-11 is a simple diagram showing a hybrid relay with solid-state control. The input relay is used to limit the voltage on the solid-state components and provide isolation between the two sections. Figure 17-12 is a similar circuit that has all solid-state components. If the output voltage is too high for the circuit, a relay can replace the load and in turn drive the higher voltage.

Figure 17-13 is another hybrid circuit only for the control of an AC voltage. Low-voltage control is again possible by using a relay. Figure 17-14 is the all-solid-state version. The use of isolation techniques is similar to those found in interfacing circuits. Isolation is a very useful feature in circuits, since specialized operations can be prevented from interfacing with other circuits, and overdriving of limited power circuits is not a problem. Arc-suppression circuits are usually incorporated into the solid-state relay to protect the transistor-type components

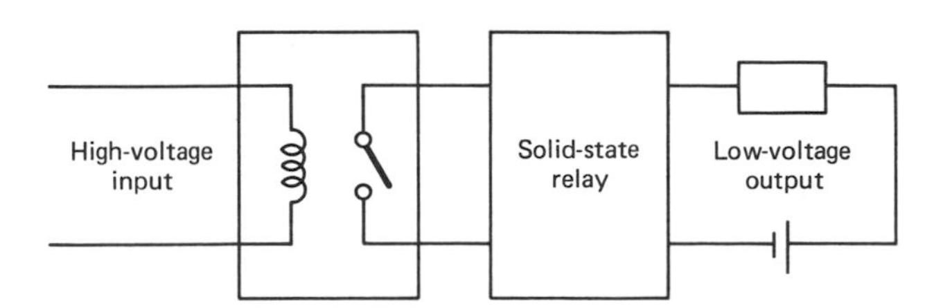

Figure 17-11 Hybrid DC relay.

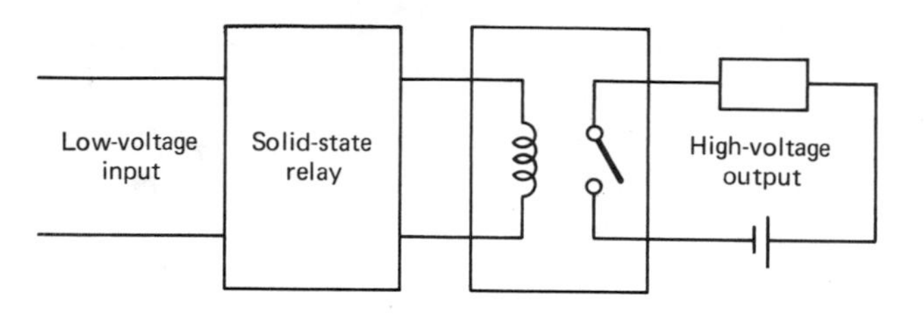

Figure 17-12 Solid-state DC relay.

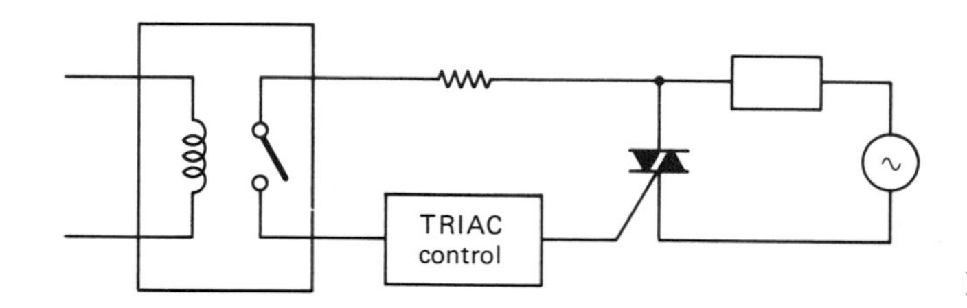

Figure 17-13 Hybrid AC relay.

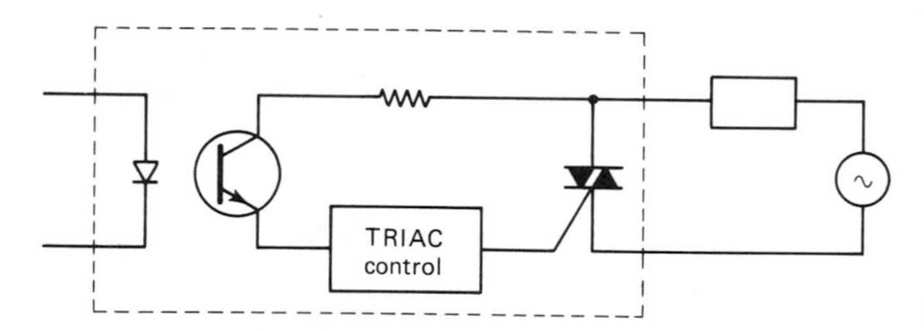

Figure 17-14 Solid-state AC relay.

from voltage and current surges which would destroy them. Application notes that are supplied with many relay products will outline the protective measures incorporated into the unit and those that need to be added by the user of the product.

SUMMARY

The relay has a wealth of research and development built into it, and this is taken for granted by many people. Knowing some of the underlying principles will not only make you more aware of proper applications but will also give you an appreciation for the expertise that is available from the manufacturer of the many products on the market. The rising cost of electrical equipment makes it even more important that you properly apply every piece of equipment you buy or install. It is much easier to explain the added cost of a relay that will prevent limited production down time than to explain why the assembly line is not running again.

EXERCISES

On a separate sheet of paper, complete a response for each question, statement, or problem listed below.

17.1. Why might a starter be considered a relay?

17.2. Identify the two main sections of a relay.

17.3. Unless specifically stated otherwise, all contacts in a relay diagram are shown in what position?

17.4. Describe the major parts of a relay.

17.5. In which portion of a relay is the force that is required to move the contacting mechanism developed?

17.6. How is the efficiency of a relay coil increased?

17.7. The strength of a relay coil is expressed in what units?

17.8. For the formula: Coil Strength $= N \times I$, what do the I and N represent?

17.9. For a relay, what effect does increasing the number of turns have?

17.10. What would happen if the wrong coil were installed in a starter?

17.11. List three of the metals that are used for making relay contacts.

17.12. Describe two features that make silver more popular for relay contacts.

17.13. To avoid problems while using relays to control input to computers, how would you reduce contact bounce?

17.14. What do the letters USASI stand for?

17.15. List one circumstance that might shorten the life expectancy of a relay.

17.16. What does MCBF stand for, and why is that information necessary?

17.17. What is the major difference between a hybrid relay and a solid-state relay?

18

Speed Controls

OBJECTIVES

Upon successful completion of this chapter, you should be able to

(1) describe the basic premise of speed control
(2) discuss the operation of common DC speed control techniques, such as SCR-controlled, manual-controlled, pulsing circuits, and amplidyne generator control
(3) describe the operation of common AC speed control techniques
(4) select appropriate AC or DC speed control circuits for different typical applications
(5) describe the advantages or disadvantages of common speed control processes

DC SPEED CONTROLS

Many variations of speed control are possible with DC motors. While we have selected several newer types of circuits, we have also selected several of the older types of circuits, since many of them still remain. State-of-the-art equipment is not always available or desirable, and for this reason it might be important for you to understand older control-item concepts.

Manually Controlled

One of the oldest methods of controlling the speed of DC motors is by the use of manually controlled resistances. The series-wound, shunt-wound, and compound-wound motor's operation has previously been examined. We will not repeat their operation but will show how a variable resistor can be added to each circuit to control the speed. The series-wound motor schematic with a variable resistor is illustrated in Figure 18-1. By varying the resistance, the amount of current allowed to flow to the armature is increased or decreased and thus controls the speed of the motor. (This is a simplified diagram and does not show all the features that are normally incorporated into this circuit.) Figure 18-2 is an example of the added features that could be included for safety reasons.

1. First the control wiper arm is spring loaded to the OFF position.
2. The second item is the coil that holds the wiper arm in place when the magnetic field of the coil is strong enough to overcome the tension of the return spring.

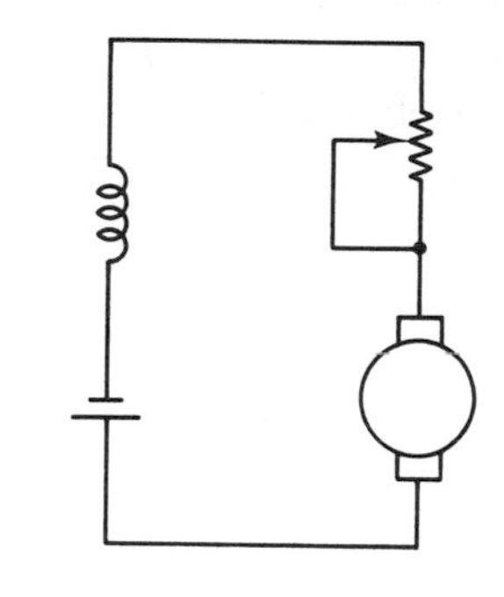

Figure 18-1 Series-wound speed control.

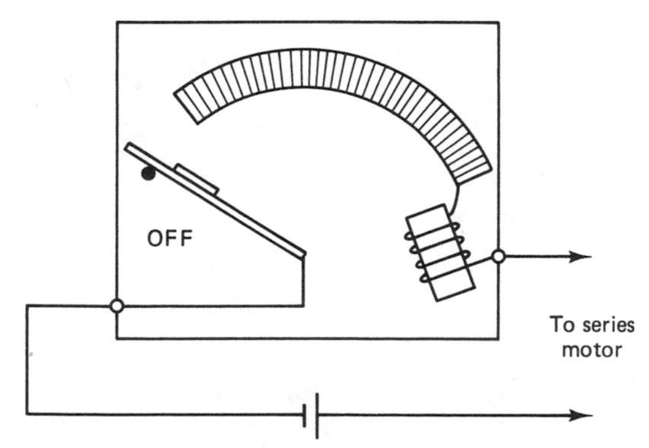

Figure 18-2 Manual speed controller.

3. The arm is at maximum resistance when the motor is started—this allows the motor to start at slow speed and gradually increase to full speed when the resistance reaches its minimum value.
4. As the current increases in the motor armature, it also increases in the holding coil.
5. After the holding coil comes in contact with the wiper arm, manual pressure is no longer required.
6. If the power is turned off for any reason, the holding coil will be de-energized and the wiper arm will return to the OFF position. This prevents it from restarting when power is restored.

This starter allows the operator to increase speed gradually but does not allow any speed to be maintained without manual control. Other versions that are not required to operate large motors may have the simple resistor control in Figure 18-1 without any spring return. This allows a speed to be selected without any manual effort to maintain it.

Figure 18-3 shows what is commonly called a *three-point starter* connected in a schematic diagram. This starter and the one in Figure 18-4 may be used for either a shunt or compound DC motor. The starters get their names from the number of wires that must be connected to them for normal operation. A second resistor in the shunt field allows additional speed control when the starting resistance is at a minimum value.

Speed control is often thought of for increasing and decreasing speed, but

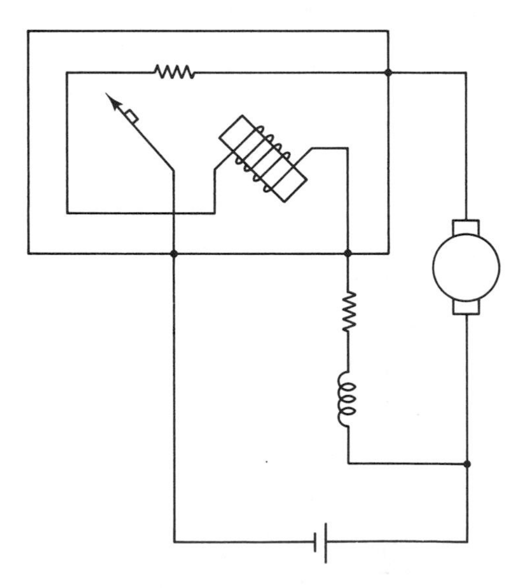

Figure 18-3 Three-point DC starter.

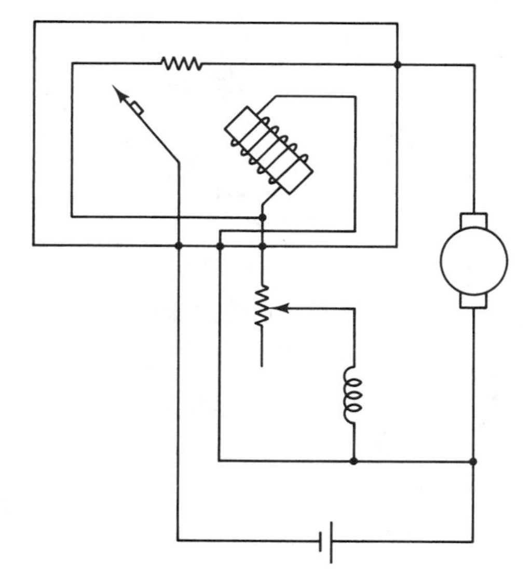

Figure 18-4 Four-point DC starter.

many starters go to the OFF position and allow the motor to coast to a stop. It is not always desirable to wait for a long period of time for the motor to stop. By adding a resistor, as shown in Figure 18-5, the motor itself acts like a generator, and the energy is dissipated in the resistor, helping to bring it to a quick stop. Figure 18-5a is a manual brake circuit, and Figure 18-5b is one possible relay version that might also be used.

SCR Controlled

Newer control circuits incorporate the use of SCRs. This is an improvement over the manual controls, since the voltage drop across the rheostat is not affected by changing load conditions. Other advantages are that smaller wattage resistors may be used in the control circuit and placed at greater distances away from the control panel if necessary. Figure 18-6 is a simple diagram of the basic components necessary for this type of circuit.

SCR-control circuits may also incorporate speed regulation when load changes occur. This can be very important with some machining operations. Figure 18-7 illustrates a typical output from an SCR-control circuit. The pulsating DC voltage enables the SCR to alter the average value of voltage and current available to the motor. Figure 18-7a shows full voltage available for the motor. Figure 18-7b has the SCR firing and reducing the voltage to near half the full value. The final waveform in Figure 18-7c reduces the voltage to nearly zero, and the motor will run at a very low speed.

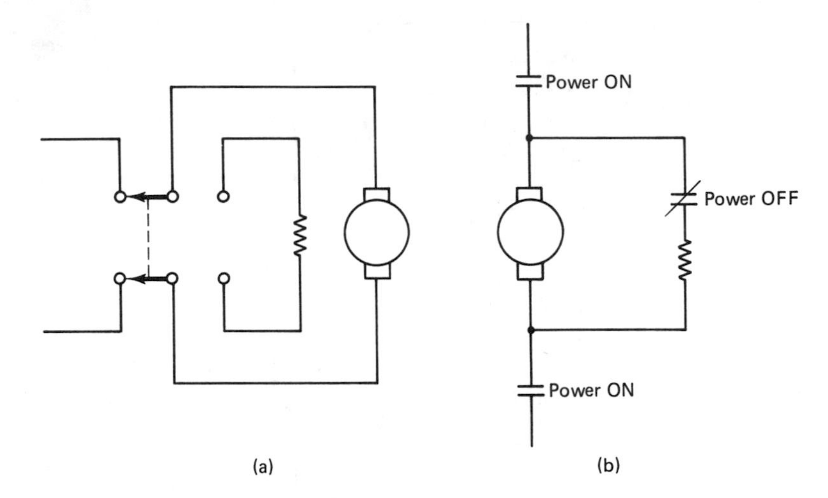

Figure 18-5 Dynamic braking.

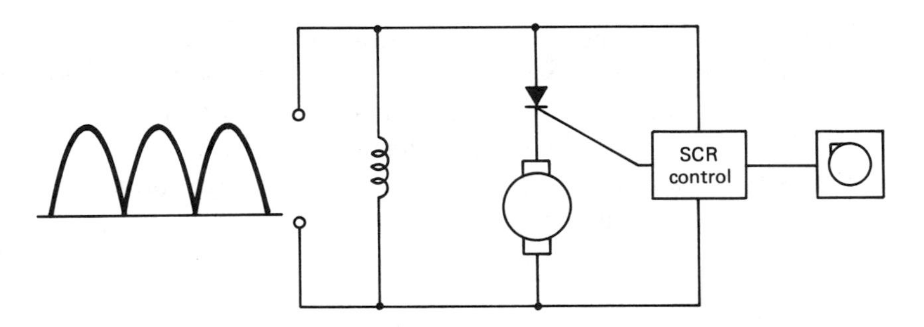

Figure 18-6 SCR DC controller circuit.

Pulsing Circuits

A similar type of control is the use of pulsing circuits. Figure 18-8 provides an example of this type of output to the motor. Instead of reducing the pulsating DC input to reduce the average voltage value, a DC voltage (battery-type source) is used, and the average value is controlled by pulsing it on and off. This creates a square-wave signal that may vary in frequency and pulse width. Increasing the frequency causes the average value of the pure DC signal to drop. The same effect can also be achieved by shortening the pulse width. Either method might be satisfactory by itself, but combining both can provide improved control.

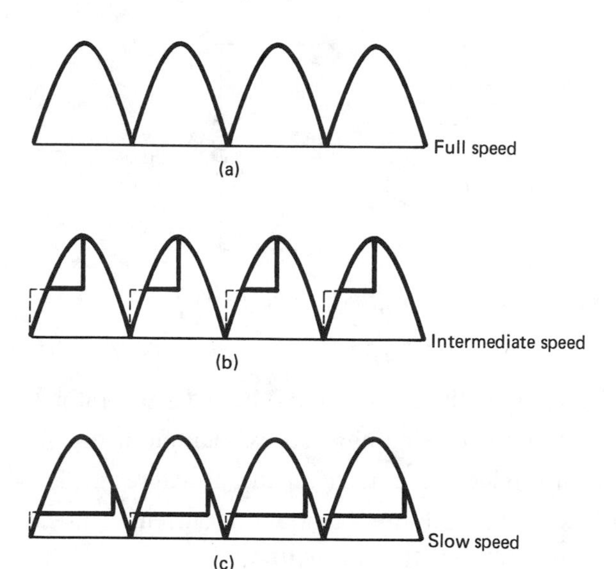

Figure 18-7 SCR DC controller outputs.

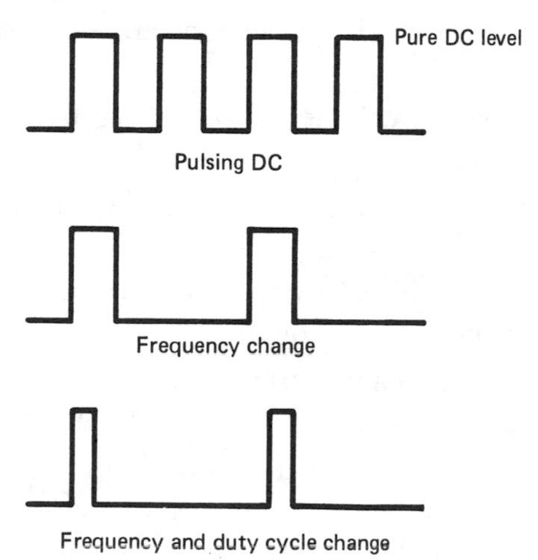

Figure 18-8 Pulsing speed control.

Amplidyne Generator Control

An interesting type of DC motor is the amplidyne. It is being replaced by solid-state controllers, but its operation remains important to those that still have them in their plants. It is possible you may never encounter one of these motors. An understanding of the operation could still be helpful to you, and it will also give you an appreciation for the benefits derived by using solid state controls. First, we will discuss the normal operation of the amplidyne which is shown in Figure 18-9.

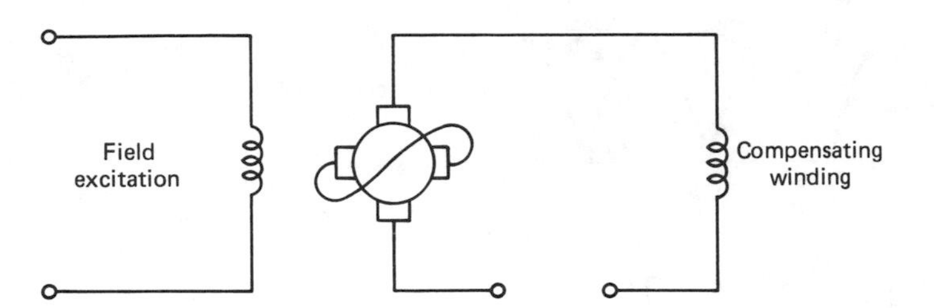

Figure 18-9 Amplidyne schematic.

1. Normal operation begins as with any other typical DC generator.
2. Field current applied to the field windings causes rotation of the armature.
3. Rotation of the armature induces a current in the armature.
4. Due to the size of the windings, a limited amount of current can be allowed to flow through them without destroying them.
5. By adjusting the field current, the speed of the armature and the amount of induced current can be controlled to protect the armature. A maximum-rated field current is also specified.
6. Instead of placing a load across the armature, the leads are intentionally short circuited as shown.
7. The ratio of the short-circuit current to the field current is called the amplifier gain.
8. To make the amplidyne useful, a second set of windings are taken at ninety degrees to the short-circuited windings.
9. The magnetic field of the rotating armature induces a voltage into these windings as well, but the magnetic field opposes the field current.
10. By adding a set of compensating windings, this field is nulled out of the circuit and only the load current remains operating at the maximum output of the armature with full field current.
11. A DC generator is physically attached to the same shaft as the amplidyne and an AC generator, which is the primary. Speed control of this load is the significant part of the unit. This is shown in Figure 18-10.
12. A smaller generator supplies a reference voltage which will oppose any increase in field current that the amplidyne may experience.
13. A decrease in field current below the reference will be aided by the reference voltage, and the amplidyne will speed up again.
14. The reference voltage is adjusted by a rheostat.

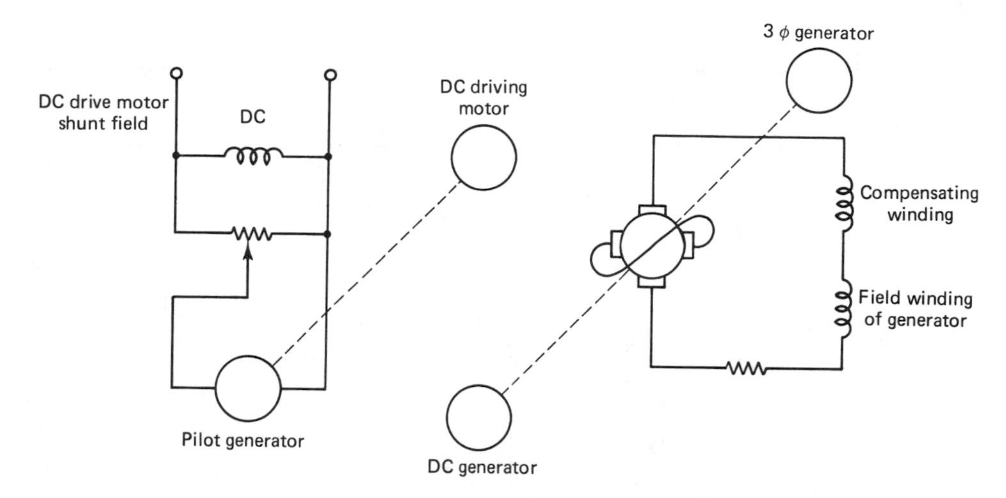

Figure 18-10 Amplidyne speed control block diagram.

AC SPEED CONTROLS

Speed control of AC motors is just as important as DC controls. The availability of solid-state control circuits has made it easier to create these circuits. It also eliminates the need for large DC power sources in many cases, As with our discussion of DC controls, we will provide examples of both the older and newer varieties.

Sequential Control

Sequential control of speeds can be accomplished by using relays and resistors as we discussed with wound-rotor motors. Such control may be done with manual switching, programmable controllers, or microprocessors. This is most important during the starting of large motors.

Multispeed Motor Control

A unique circuit is presented in Figure 18-11. This circuit shows only the power connections of a circuit used to change speeds on a multispeed motor. A schematic diagram of the motor connections and the nameplate information required to change speeds is also illustrated. The nameplate data could be used to wire the motor for a single speed, but by using a series of motor starters, different speeds

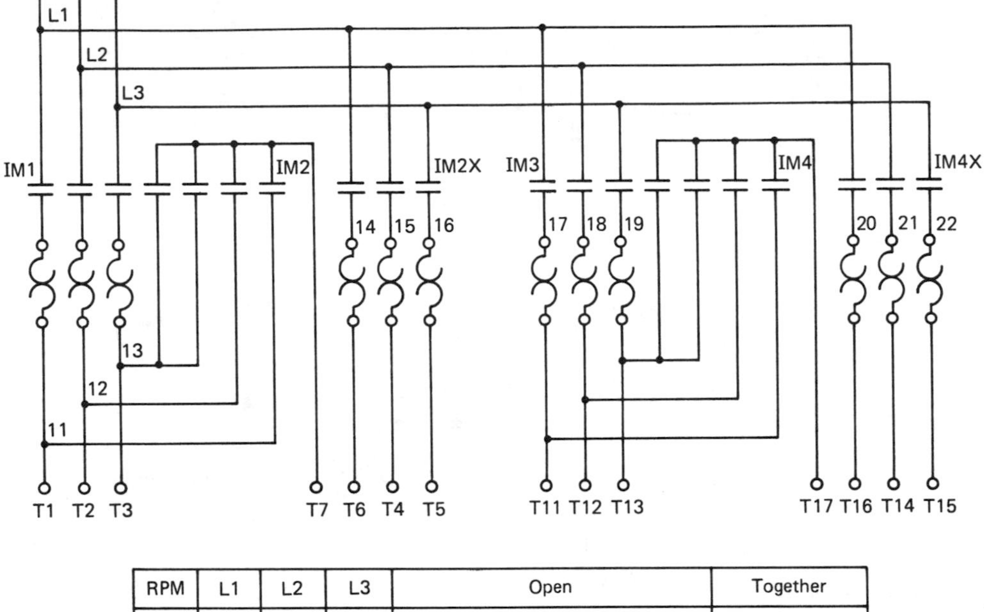

RPM	L1	L2	L3	Open	Together
450	T1	T2	T3, 7	T4, 5, 6, 11, 12, 13, 14, 15, 16, 17	—
900	T6	T4	T5	T11, 12, 13, 15, 16, 17	T1, 2, 3, 7
1800	T11	T12	T13, 17	T1, 2, 3, 4, 5, 6, 7, 14, 15, 16	—
3600	T16	T14	T15	T1, 2, 3, 4, 5, 6, 7	T11, 12, 13, 17

Figure 18-11 Multispeed motor.

can be set by using a selector switch (not shown). Refer to both the schematic diagram of the motor and the nameplate information as we step through the speed selections.

1. Speed one is selected when starter M1 is energized.
2. The four contacts of M1 close and power is available at T1, T2, T3, and T7. All other starters are de-energized; now power is connected to the load side of these starters.
3. Notice that starter 1M2 is open. If this starter were to close at this time, all three phases would be shorted together.
4. Figure 18-12 is the final circuit connection on the motor for speed one.
5. When speed two is selected, starter M1 is de-energized, and starters 1M2 and 2M2 are energized. All other starters are de-energized.
6. Starter 2M2 connects power to T4, T5, and T6.

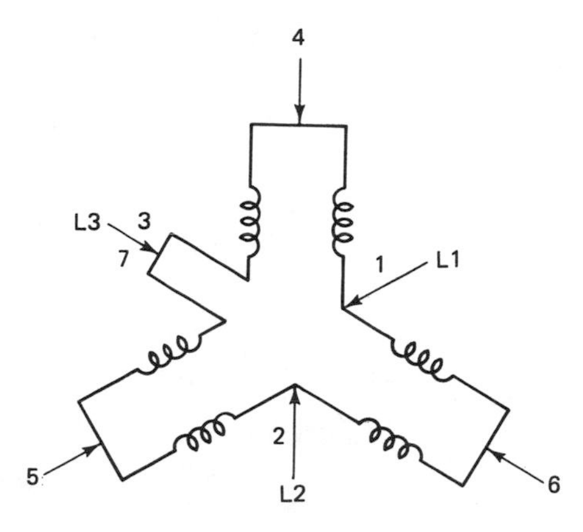

Figure 18-12 Motor connections speed one.

7. Starter 1M2 shorts leads T1, T2, T3, and T7 together. This would short out the power supply, but since M1 is de-energized, no power is present at terminals T1, T2, or T3.
8. Figure 18-13 shows the motor connections for speed two.
9. The windings are placed in parallel—this allows more current flow to the motor. It speeds up.
10. When speed three is selected, starter M3 is energized and all others are de-energized.
11. The four contacts of M3 connect power to T11, T12, T13, and T17.
12. Starter 1M4 is de-energized—this prevents short-circuiting of the power connections.
13. The circuit diagram for speed three is shown in Figure 18-14. Notice the similarity to that of speed one.

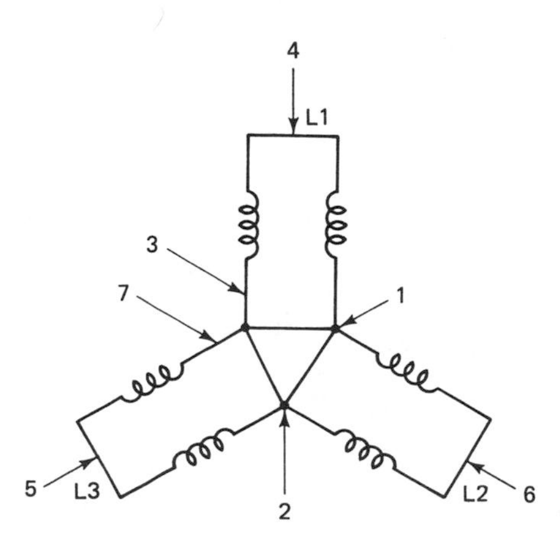

Figure 18-13 Motor connections speed two.

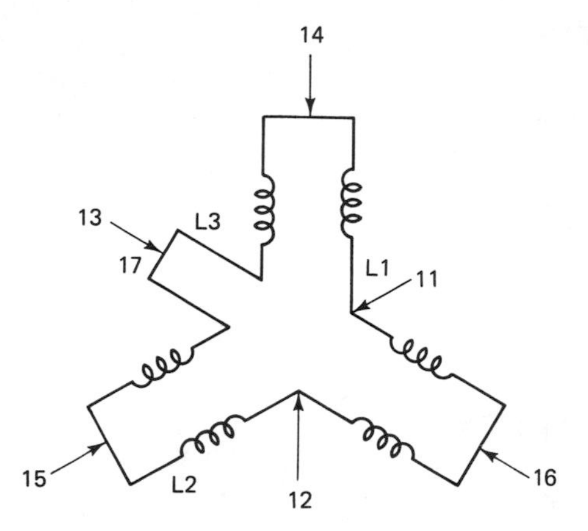

Figure 18-14 Motor connections speed three.

14. The resistance of this circuit is less than that of speed two and current increases, speeding up the motor once more.
15. When speed four is selected, all starters now de-energize except 1M4 and 2M4.
16. Starter 1M4 connects leads T11, T12, T13, and T17 together.
17. Starter 2M4 supplies power to leads T14, T15, and T16.
18. Figure 18-15 shows the final circuit diagram for speed four.
19. Once again the parallel arrangement allows more current to flow than in speed three, so the motor again speeds up.

The control diagram was not given for this circuit, since it is quite involved. The idea is for you to see how starters may be used to change speeds in the same motor. Many interlocks are required in the control circuit to prevent accidental closing of some contactors. We hope you also noticed how going from a wye to a delta arrangement might be accomplished with starters.

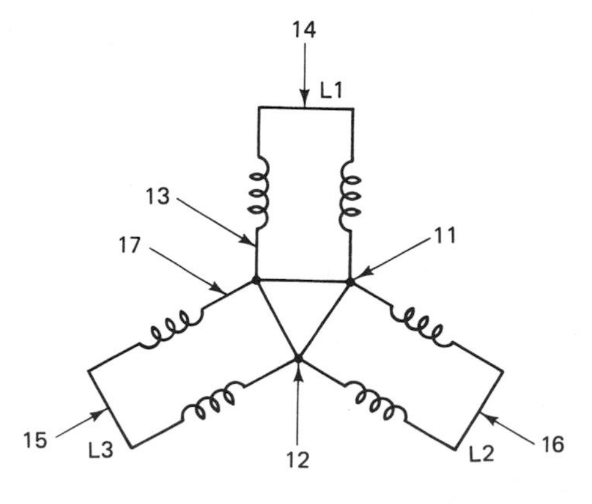

Figure 18-15 Motor connections speed four.

Alternating Current SCR Speed Control

Motors that require AC power sources can also be controlled by using SCRs. Their operation is similar to that of DC motors. SCRs are used to reduce the effective operating frequency of the motor. Low-frequency signals cause the motor to slow down, and high frequencies cause it to speed up. Circuits that control these motors often have features to determine loading effects and the rate at which the load is changing. The circuit then compensates for these changes before the motor is allowed to change speed.

NUMERICAL CONTROL

As systems become more automated, the machine is becoming the major controlling device. Programmable controllers and microprocessors are monitoring functions making the necessary corrections to keep the system within design tolerances. One type is numerical control. Numerical control was one of the first methods of programming the entire operation of a machine by using numerical codes. These codes were placed on paper tape and fed through a reader which interpreted each group of holes and then caused the machine to proceed to a specific operation. In addition to having relays change the speed of the motor, they may also reposition parts, or raise and lower a drill press, for example. Precise speeds are necessary for drilling and milling operations to prevent the tool that is being used from overheating or breaking. To obtain the speed of the motor, a tachometer or tach-generator is incorporated so it can send data back to a control circuit, which in turn indicates rotation of the motor.

The tach-generator, as the name implies, functions as a voltage generator. Let's briefly examine how it functions. Figure 18-16 is a block diagram of this control system.

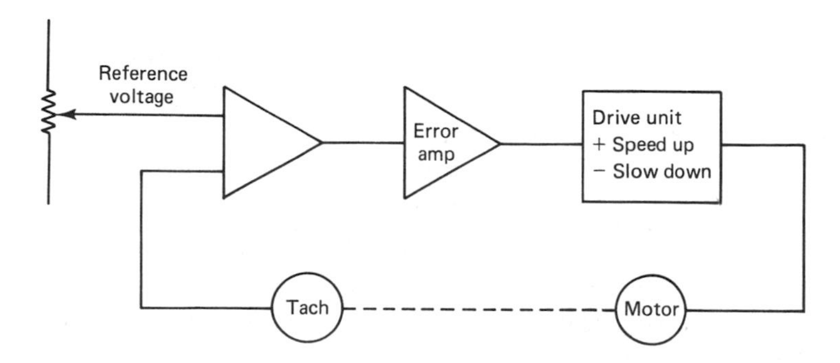

Figure 18-16 Tach-generator control.

1. A voltage reference is established in the control circuit that represents the speed of the motor from zero volts at zero speed to some maximum voltage level at maximum speed. When the tape is programmed, the reference speed is set.
2. A shaft from the tach-generator is physically connected to the motor shaft. The tach-generator therefore rotates at the same speed as the drive motor.
3. A voltage that is proportional to motor speed is developed in the tach-generator and sensed in the control circuit.
4. If the generated voltage and the reference voltage are the same, then no error exists and no corrective action is required.
5. A second circuit determines the exact amount or type of error and sends a signal to the appropriate circuit either to increase or decrease motor speed.
6. Assuming the motor was running too slowly and the sensing circuit sent the correct signal to the speed control section (maybe an SCR speed-control circuit), the motor should then increase in speed.
7. At some point in time the voltage generated by the tach-generator will equal the reference voltage. At this time, zero error exists, and the no-correction voltages are developed.
8. Any further change in motor speed that causes an error signal will also cause a correction signal to be developed.

What we have just described is a feedback control system. This is not unique to numerical control, but it is a very important method of speed control. One of the limitations of numerical control is that the machine will follow only the instructions on the tape. Manual interruption is required to change any process that has not been coded into the machine, or a new program tape must be made. If the machine is very reliable and problems rarely develop, the numerical control system is an excellent method to provide many repetitive operations with high tolerances. Even the most reliable machine still requires some human intervention. This might be changing a program tape, developing a new program, changing a tool after a series of operations, calibration or other similar tasks. When such a machine does break down, it will be up to someone with some knowledge of controls to repair it.

Related to numerical control systems are the newer microprocessor systems. The program that was placed on a paper tape can also be coded onto a magnetic tape or disk. Instead of reading each code in the paper tape as it proceeds through the reader, sections or the entire operation code may be loaded into the memory of a microprocessor. The process can also be monitored at each of the program steps and corrections made when required.

Recall the sampling cycle we examined with programmable controllers. Microprocessors that operate at high speeds and sample data at short intervals can send information to many circuits. Speed control can be affected not only by the tach-generator and motor speed but also when the more advanced machine might sense excessive wear on a drill bit and turn the machine off, remove the bit, pick up a new one, and resume the normal sequence of operations. Microprocessors also allow changes in the programs without interrupting normal operations and allow monitoring of the complete cycle from locations other than at the machine. Data on the number of parts made, time required for each step or cycle, and time and reasons for failures can be generated in graphic or paper reports, as required.

OTHER CONTROL FACTORS

There are times when we get overly involved in technological advances and overlook some of the more practical solutions to problems. Speed control can be one of these areas. You should learn to solve control problems with the means that will provide not only the desired results but also incorporate safety (always first), the simplest method, machine limitations, and also budget constraints. Buying a thousand-dollar speed-control unit when a ten-dollar pulley would provide the desired results, is a good example. Two factors that should be considered are pulley (or gear) size and motor base speed. The first factor we will examine is pulley size.

Figure 18-17 has the same size pulley on both the motor and the load. The speed of the motor and load will be the same. The speed of the load is proportional to the ratio of the drive pulley to the load pulley.

Figure 18-18 illustrates the condition when the load pulley is both greater and smaller than the drive pulley. A similar situation occurs with gear driven machines, as shown in Figure 18-19. As the speed ratio is changed, torque is also altered. In general, when the speed is reduced the torque is increased.

Selecting a motor with a different speed may also provide the necessary speed control or change that is desired. This may not be the best alternative in all cases, but it still remains an option available to you. These are fixed-speed

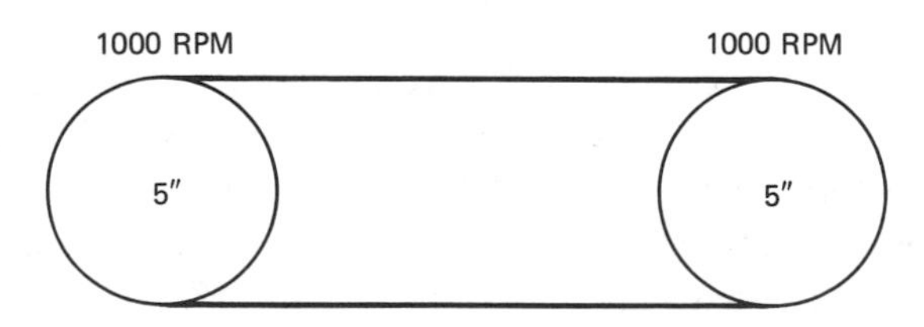

Figure 18-17 Pulleys with 1 : 1 ratio.

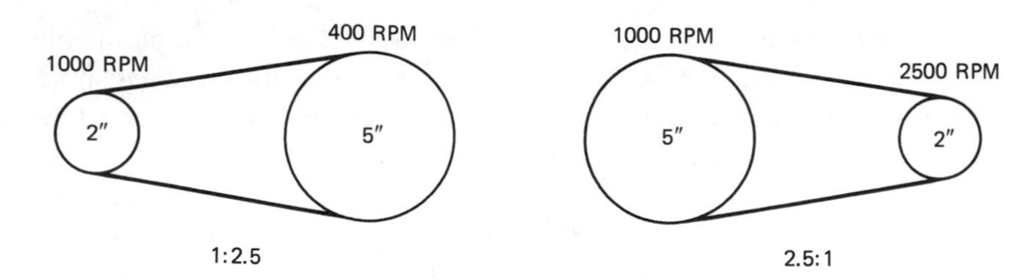

Figure 18-18 Ratios other than 1 : 1.

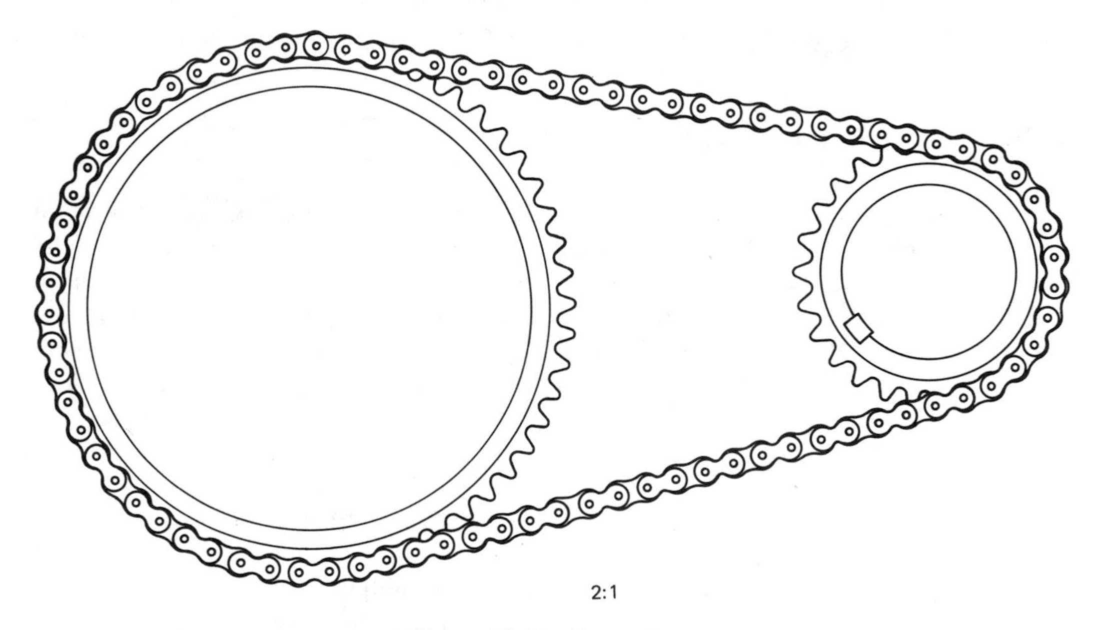

Figure 18-19 Gear ratios.

solutions to the problem. Electronic speed controls or variable speed motors are the only other options that can provide speed changes while the motor is operating.

SUMMARY

Speed control is important in many circuits. For several reasons the large number of variations available were featured here. The emphasis of both old and new versions was given to allow those who still use these devices to maintain them and allow for transition to newer equipment as it becomes available. Safety and common sense are also an important part of any control circuit—hopefully you will incorporate this into your thinking as well.

EXERCISES

On a separate sheet of paper, complete a response for each question, statement, or problem listed below.

18.1. For manual speed control, which component is most commonly used?

18.2. What advantage does an SCR control have over manual control?

18.3. What is the purpose of the pulsating DC voltage in an SCR control?

18.4. How does the pulsing circuit achieve its control function?

18.5. When using amplidyne generator control, what do we refer to as the amplifier gain?

18.6. What is a major advantage of the AC speed control over the DC type?

18.7. List the three most common devices that are used for sequential control.

18.8. For a multispeed motor control, why is the data listed on the nameplate important?

18.9. For a multispeed motor control, why are many interlocks necessary for the control circuit?

18.10. For an alternating current SCR speed control, when does the control circuit compensate for load changes and loading effects?

18.11. What is numerical control?

18.12. For drilling and milling operations using numerical control, why is precise speed necessary?

18.13. List a major limitation of the numerical control.

18.14. In prioritizing the factors you must consider before purchasing a piece of equipment, where would you place safety?

19

Control Systems

OBJECTIVES

Upon successful completion of this chapter, you should be able to

(1) differentiate between open-loop systems and closed-loop systems
(2) differentiate between proportional control, ratio control, and reset control
(3) describe the operation of the on-off control
(4) connect typical circuit configurations that incorporate pressure sensors, flow meters, temperature sensors, and radiation measurement sensors

OPEN-LOOP SYSTEMS

Control systems are usually classified as either open-loop or closed-loop systems. The choice depends on the application of the system, the degree of sensitivity, the desired outcome, and the cost effectiveness of equipment used to control a process. Graphically, the open-loop system is like a one-way street. A control signal is sent to the controlled device; no further interaction occurs until the signal is removed. On and off applications are the best example of this.

A simple light switch is an open-loop system. The block diagram in Figure 19-1 shows the one-way relationship between the signal and the control device. This circuit provides the least amount of interaction between a signal source and the equipment being operated.

CLOSED-LOOP SYSTEMS

The closed-loop system adds at least one new important element—feedback. We briefly gave an example of feedback for the speed control circuit in the last chapter. We need to examine this more closely. Figure 19-2 illustrates a simple feedback block diagram. A second requirement is a summing network that combines an input signal with the feedback signal, to indicate the amount, or degree, of control desired for the controlled device. By providing a polarity to the feedback signal, the summing network can either increase or decrease the magnitude of the final output signal. This effect can be seen in Figure 19-3 for negative feedback and Figure 19-4 for positive feedback.

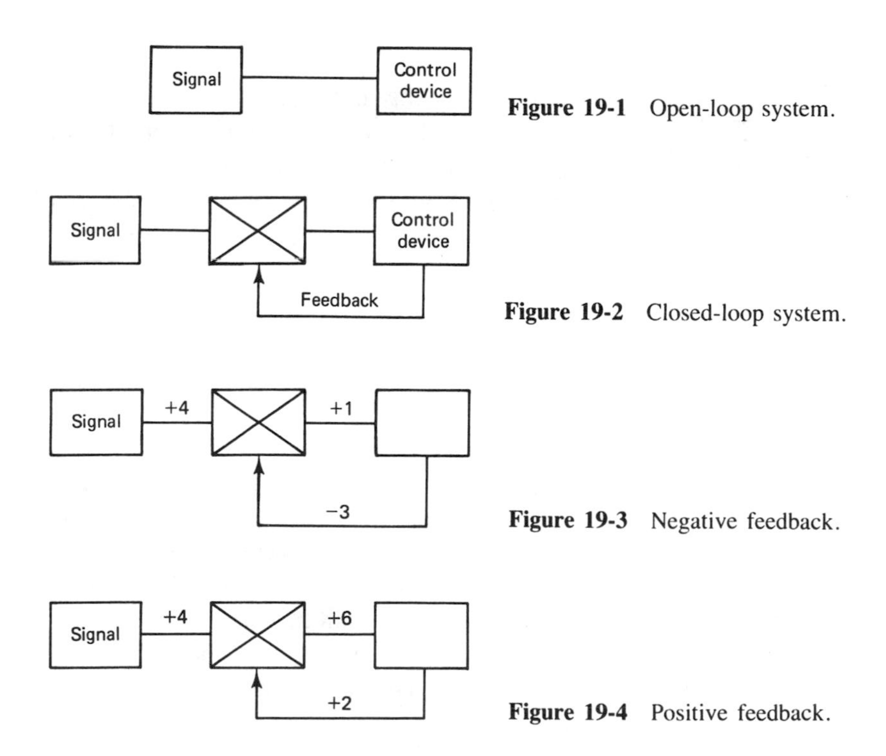

Figure 19-1 Open-loop system.

Figure 19-2 Closed-loop system.

Figure 19-3 Negative feedback.

Figure 19-4 Positive feedback.

1. The signal source is sending an output of +4 to the summing network.
2. Without the summing network, the final control device will move to a position that corresponds to a +4 signal.
3. The final element in this case is sending a negative feedback signal to indicate a need to reduce the input by a level of three (a negative sign represents negative feedback).
4. The summing network combines both of the signals. The resultant is an output with a +1 magnitude. This is the algebraic sum of the two inputs to the summing network.
5. When the value of the feedback signal is equal in magnitude and opposite in polarity, the output signal will be zero. This is the desired condition in a closed-loop circuit. This indicates that no correction is needed to obtain the desired value or position of the controlled device.
6. Figure 19-4 shows a condition where the output is determined to require a +4 level of correction, but the controlled element is actually at a position greater than +4 away from the zero or desired condition.
7. The +2 level (positive feedback) indicates this condition.
8. The additional correction is supplied by the summing network as a +6 level output. This will drive the controlled element toward a zero or null condition.

One of the reasons for using motion as an example is that in many instances open-loop signals are utilized for position control. Let's look at an example using a DC motor connected to a feedback potentiometer. A bridge circuit will be used to develop the output signal. Figure 19-5 shows a DC servomotor connected to a damper. The dotted lines to the motor and potentiometer R4 represent mechanical connections that move when the motor is in motion. Resistor R1 is the input control to the circuit. Points A and B are where the output is developed across the motor. The polarity of the output determines the direction of rotation of the motor. Before we examine this circuit, it might be helpful to review the operation of a bridge circuit.

Figure 19-6 is a simple bridge circuit. The basic theory of operation suggests that the potential difference across points A and B will be determined by the ratio of resistances in the circuit. When the ratio of R1 to R2 is equal to the ratio of R3 to R4, then the potential difference between these points will be zero and no current will flow. If the ratio of one set of resistors is changed for any reason, and the other set is not adjusted to maintain the same ratio, then a potential difference will exist. The actual value of each resistor is not as important as the ratio that exists. Try proving this to yourself by taking the milliammeter out of the circuit

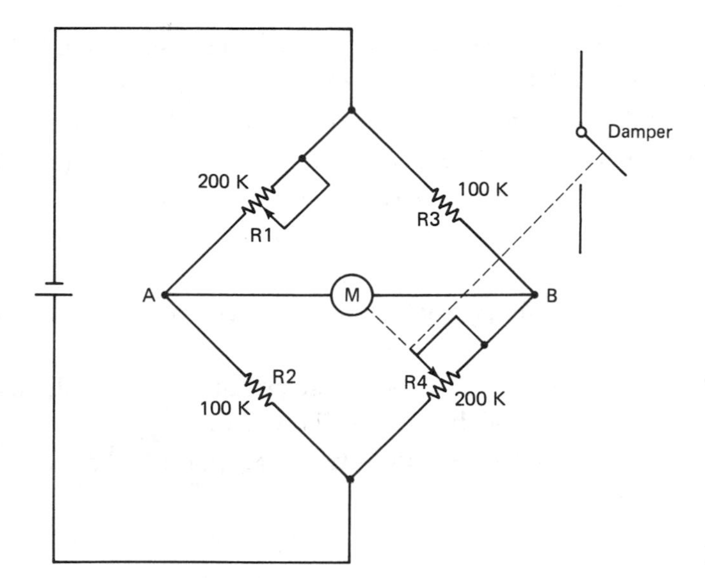

Figure 19-5 Servomotor circuit.

and calculating the voltage at both points A and B with different values of resistance. With this theory in mind let's discuss the operation of the circuit in Figure 19-5.

1. The first condition is where points A and B are at the same potential.
2. No current flows and the motor does not move.
3. This implies that the ratio of R1:R2 is equal to R3:R4. Assume all the resistors are set to have equal values.
4. The position of the damper is adjusted for a zero position with the circuit at initial set up. Assume this is as shown in the diagram, with the damper half open.

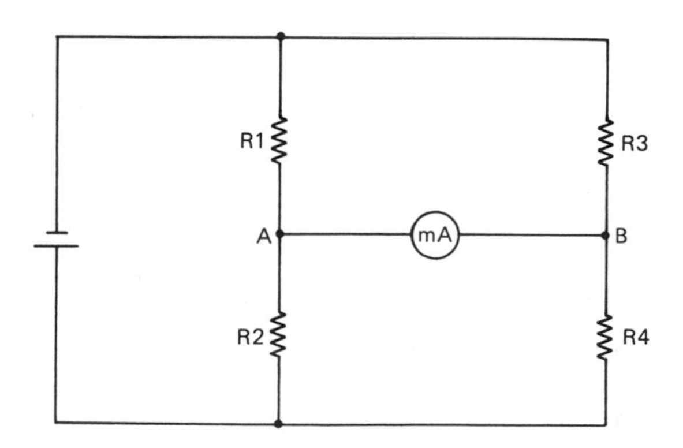

Figure 19-6 Bridge circuit.

5. Moving potentiometer R1 away from point A increases the resistance of R1. This increases the ratio of R1:R2.
6. The voltage at point A will then have a greater magnitude than at point B.
7. A positive potential at point A with respect to point B will cause current to flow through the motor, and it will rotate.
8. The shaft of potentiometer R4 is connected to the motor and will move as the motor rotates.
9. If the wiper arm of R4 moves away from point B, the ratio of R3:R4 is also increased.
10. At some point the ratios of both sides of the bridge circuit will be equal, and the voltage drops at points A and B will also be equal.
11. No potential difference exists between the two connections of the motor, and it no longer rotates.
12. Since the damper is also attached to the motor, it will close in this example.
13. By adjusting R1 in the opposite direction, the operation will reverse itself and the damper will open.

The preceding illustrates basic operation of a feedback circuit to control the position of a damper. Although it does not physically coincide with our block diagram of the closed-loop system, it does fulfill the basic needs of having an input signal, output signal, summing network, and feedback circuit. This is a very useful circuit, but it does have some limitations.

CLOSED-LOOP SYSTEM REFINEMENTS—TYPES OF CONTROL

Our damper circuit seems to be functioning without any problems as we described it, but there are a few hidden areas of concern that need to be examined. Various types of control are used to compensate for different control situations. Some of the names given to these types of control are on-off, proportional, and rate control.

On-off Control

On-off control is simply turning the control circuit on, letting it go to the full limit of operation and turning it off. Correction for the new condition is done in the same manner. The control element may travel from one operational limit to the other. This does not allow fine control that is needed for some control elements. Figure 19-7 shows the range of operation for on-off control. The rate of change is shown as a sawtooth waveform, since the signal may be full on or off, but the rate of change is still determined by the physical limits of all the equipment used. This includes the speed of the motor and the size of the control levers, as well as the location of the fulcrum on the lever.

Proportional Control

Proportional control allows the magnitude of the output signal to be proportional to the difference in the correction required. If the damper were fully closed and it needed to be opened to the 80% level, the initial speed of the correction would be rapid. As the difference between the desired level and the actual level decreases, the rate of correction also slows. This circuit should prevent the rapid overshoot that is apparent with the on-off control. Some overshoot does exist at the beginning of the cycle, but it gradually dampens out, since only small corrections are necessary after a period of time. This system works well when changes are rapid, but not frequent. It needs time for dampening to occur. If this time is not available, the system will continue to search for the desired output level but will be unable to level out before the next large change occurs. It will continue to hunt for the desired position. Figure 19-8 illustrates the rapid correction and dampening effect of proportional control.

Rate Control

Rate control circuits monitor the rate at which the signal is changing and make corrections accordingly. If a rapid change is sensed, then a rapid correction is made. Slow changes have signals of lower amplitude which allow gradual changes.

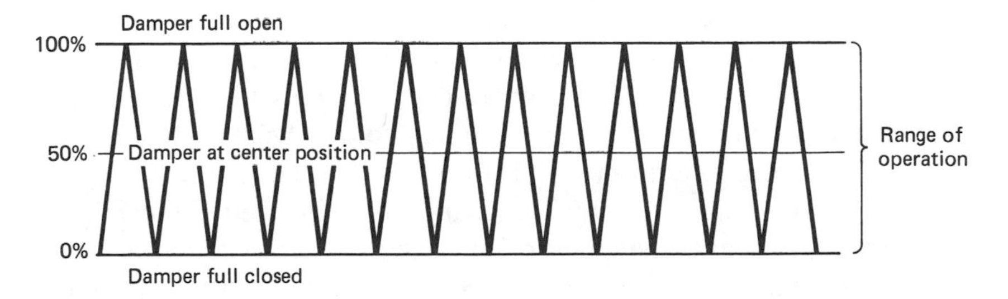

Figure 19-7 On-off control range.

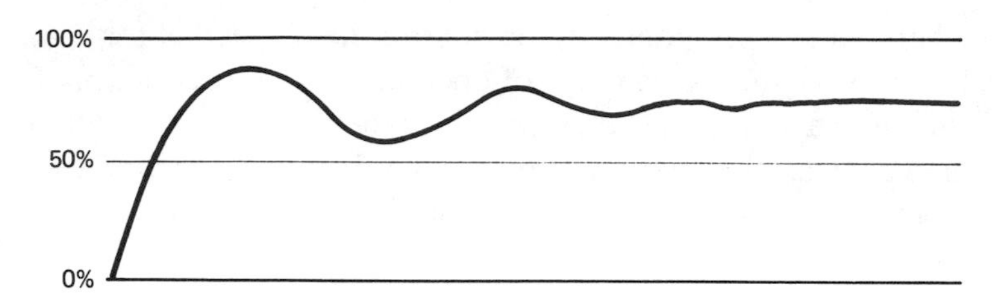

Figure 19-8 Proportional control.

Reset Control

One final type of control that may be used is *reset control.* When processes require set points of operation, the controller is calibrated to operate at the desired output or setpoints. The controller then continuously inches the output toward the setpoint. If the output is above the setpoint, a small input is always working towards it, and in this way the controller can anticipate the polarity of the output but not necessarily the magnitude of the correction required.

It should be emphasized that any or all of these circuits may be added to a control circuit to provide the desired output. The range of operation may also be changed. The type of process will determine which system is best.

SENSORS

To have an effective control system, the correct type of sensing unit must be used. There are many different types of sensing units available, even for the same purpose. Generally these units are called transducers. Although the name implies that some signal is transformed (sensed) and sent to a receiver unit to process the signal, it does not describe the method by which this is done. The name transducer means to transform or change. If we have an electronic controller and the transducer is measuring water level, there must be a means of converting the level measured into an electrical signal. This is done with a variety of devices. Since transducers convert one type of signal to another type (such as mechanical to electrical, electrical to mechanical, pressure to current or voltage, and current or voltage to pressure), a physical change has occurred. Transmitters used to send signals may convert either current to voltage, voltage to current, current to current or voltage to voltage. The transducer is often connected to a signal transmitter to convert its output to match the system's receiving unit.

Mechanical Sensors

One method of mechanical sensing can be a simple ON-OFF switch that indicates a specific level. Another may be a variable resistance to indicate degrees of level change. Figure 19-9 shows these two methods. This will begin our examination of types of transducers. In our previous example, we measured water level, but as you soon discovered, there are several different ways of doing this. The ON-OFF control of the float switch is limited in its action, but the variable resistance can be used in many different control circuits. The resistor with a float attached is one example of a mechanical sensor. For example, a mechanical sensor takes motion or position directly and transforms it to a position on the sensor to indicate the water level.

A second type of mechanical sensor is the linear variable differential transformer or LVDT. Figure 19-10 shows a simple example of this sensor. We need to examine briefly the operation of the LVDT to understand what effect it has in our diagram. Notice that the LVDT has three separate coils. The two outer coils have greater signal magnitudes when the iron core is placed within them. When the iron core is centered, both the output coils have equal but opposite signals. Thus they null each other out to have, in effect, a zero output. When the water level is low, the core moves into the lower coil, and its output will be greater than the upper coil's. By applying this signal to the proper control circuit, a valve may be opened or closed or a pump may be turned off. It all depends on the type of control action desired. The LVDT works on the principle of changing

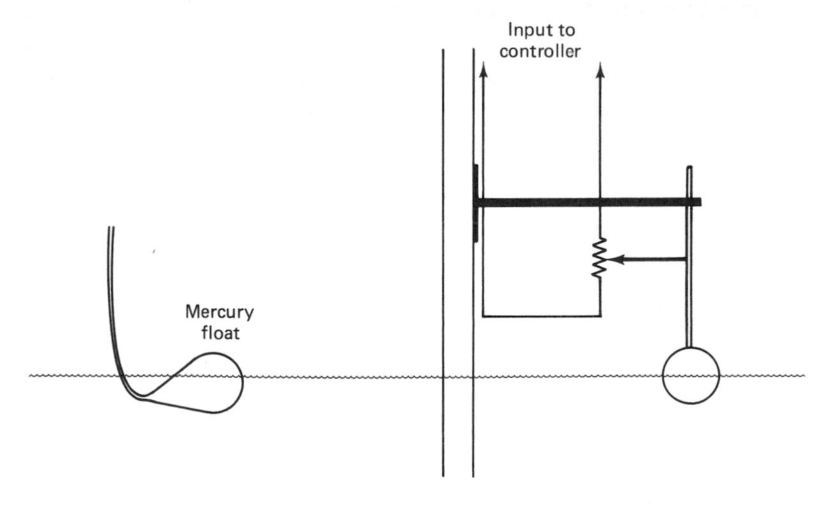

Figure 19-9 Transducers for water level.

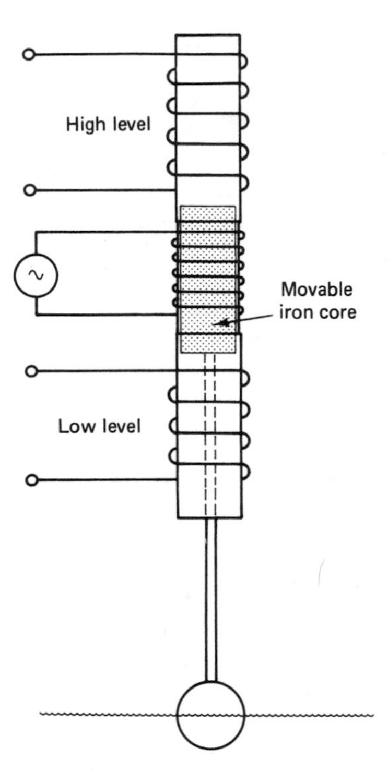

Figure 19-10 LVDT sensor.

inductance. Although we greatly oversimplified the use of the LVDT, it should be noted that very sensitive measurements are possible with them. Millimeter and micrometer values are common; so we tend to find LVDT used more for vibration detection when this is a critical factor. They are also subject to interference from the shock of someone's bumping into them or magnetic interference from radios. Our float LVDT would really not be the best way to use it, but it will aid us in comparing various methods of sensing.

Another important mechanical sensor is the strain gage. Figure 19-11 is an example of the structure of a typical strain gage. Maybe you recall some of the

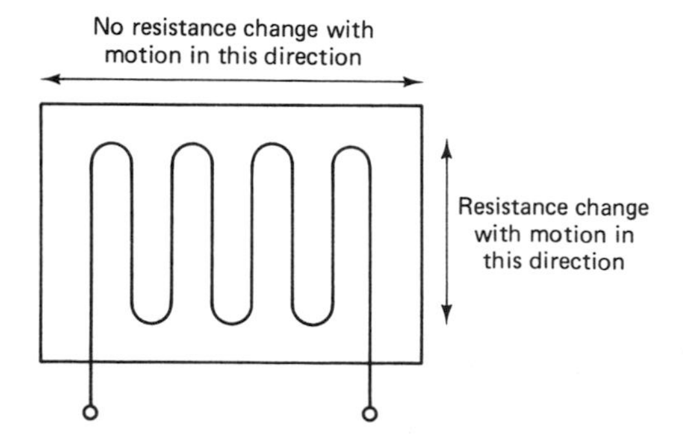

Figure 19-11 Strain gage.

resistance effects of making a wire larger, smaller, longer, or shorter. The same principle is used here. When the strain is placed on the gage in the two directions shown, you get two different effects. As long as the motion is in line with the direction in which the coils are laid out, the wire is not stretched and no resistance change takes place. By pulling in a motion ninety degrees to this axis, the wire is stretched and the resistance becomes greater. Very small resistance changes occur, so a good amplifier must be used to make the signal usable.

Pressure Sensors

Many times it is necessary to measure a pressure with a sensing device. A method of measuring water level can use changing capacitance to develop an output signal. This involves the use of a pressure transducer. Figure 19-12 illustrates how a capacitor transducer might be used to measure water level. The two plates are enclosed in a water-tight container, and a diaphragm is attached to the one plate that is movable. Higher water levels exert greater pressure on the diaphragm—this moves it closer to the opposite plate (which is similar to the effect that diving into the deep end of a swimming pool has on your ears). The deeper you go, the more pressure you feel. The changing capacitance is used in a sensing circuit to change a controlled output. This might be an increase or decrease in voltage or current that corresponds to the changing pressure.

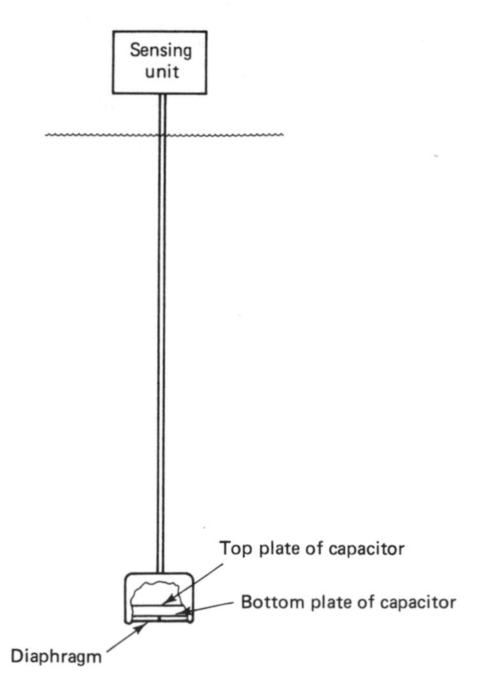

Figure 19-12 Capacitance transducer.

Pressure may also be measured by using Bourdon tube-type instruments. These tubes are coiled, and they will bend or unwind as the pressure in them is increased. Figure 19-13 shows an example of this type of meter. By attaching the wiper arm of a potentiometer to part of the coil, the increase or decrease in pressure will result in a change of resistance. This is one of the most popular types of pressure transducers. They may also have a mechanical gage to display the pressure reading.

Flow Measurement

Flow measurements are also an important characteristic to be sensed. The devices used to measure flow can be as simple as a paddle connected to a potentiometer or more complex with rotating vanes connected to pulse-sensing systems or tachometers. As flow increases, the speed of rotation of the sensing device also increases. In the case of a tachometer, the output voltage will increase or a frequency increase may be sensed. Pulse counters utilize a wheel that rotates. A part of the wheel has a magnet on it, and as it passes a magnetic-sensitive part of the circuit, a pulse is induced into it. The number of pulses are counted and converted to a flow measurement. Figure 19-14 is the top view of a pulsing pick-up unit.

Temperature Measurements

Temperature measurements are required in many processes. As with other sensing devices, we find a variety of methods available to us. Since temperature is one of the most sensed items in industry, we need to examine it a little more closely. Three temperature-sensing elements of importance are the thermistor, thermocouple, and resistance thermal detector or RTD.

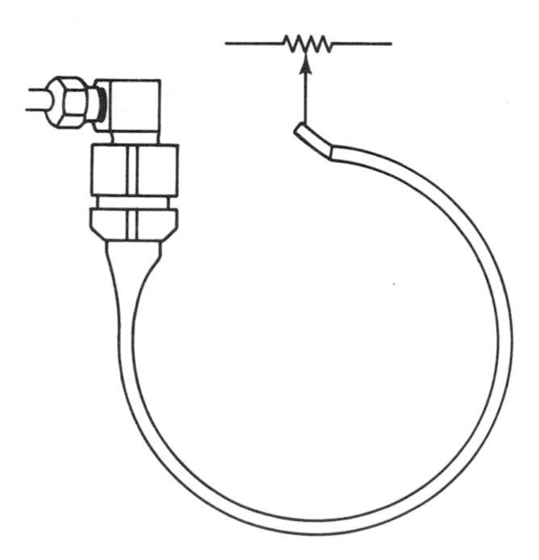

Figure 19-13 Bourdon tube.

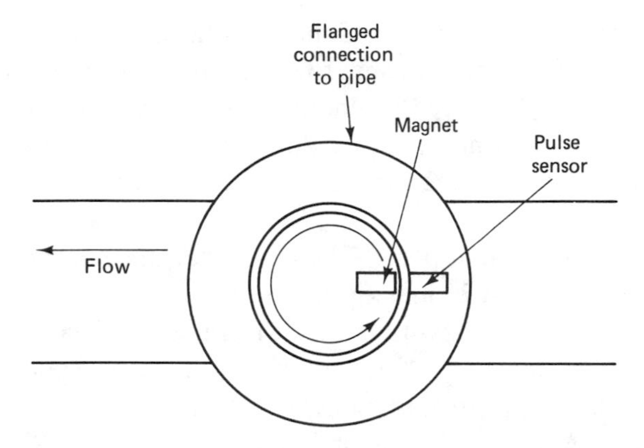

Figure 19-14 Pulsing unit.

The *thermistor* is a semiconductor device which has a negative temperature coefficient. A negative temperature coefficient indicates that resistance decreases as temperature is increased. This occurs at an exponential rate, as shown in Figure 19-15. Thermistors are not considered to be precise, but a tolerance of one half to one percent is possible, with ten to twenty percent being the norm. Resistance values of one thousand to one million ohms are typical at 25 degrees Centigrade. The thermistor is commonly placed in one leg of a bridge circuit.

Thermocouples are heat-sensitive devices made by joining two dissimilar metals together to form what is called a junction. In simple terms, when this junction is heated, it generates a small voltage. The magnitude of this voltage is proportional to the amount of heat that is applied to the junction. However, a thermocouple circuit is actually more complicated than this. First, there must be at least two junctions—a hot and cold junction. The hot junction is the measuring or sensing junction, and the cold junction is the reference junction. The thermo-

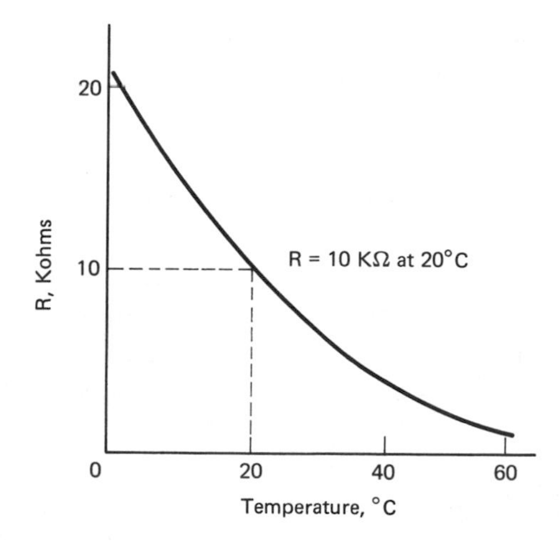

Figure 19-15 Thermistor curve.

couple circuit requires a complete loop to function, just like any other circuit. If the hot junction is heated by a flame or exposure to the element to be measured, a voltage is generated at the junction. Since the junction is made up of two different metals, current will flow with a definite polarity through the circuit. As long as the reference, or second junction, is not at the same potential as the first junction, current will continue to flow.

This brings up a second important point. The polarity of the current flow is dependent on which metals are used in making up the thermocouple. There are several different combinations of metals used. Each has its own characteristics when coupled with another conductor. Thermocouples have been standardized and four combinations of conductors are commonly used. These are the Chromel-Alumel, Iron-Constantan, Copper-Constantan, and Platinum–Platinum-Rhodium. The first metal named is the positive element of the thermocouple, and the second metal is the negative element. There are many more combinations possible. Instead of saying, for example, each time that a thermocouple is Chromel-Alumel, possible combinations were given a standardized letter. In this case it would be type K. Since most thermocouples look pretty much alike, there is also a need for some method of determining which thermocouple is of what type in the field. The conductors coming out of each thermocouple have also been standardized. The positive wire in each type is given a color code. The negative wire of all types is color-coded red. Each pair of wires also has an outer jacket that is color coded. Figure 19-16 is an example of the types and color codes used for thermocouples.

Most thermocouples have short leads terminating in a small junction box on top of them. If two copper wires are added to an existing thermocouple, then two additional junctions are created. By connecting identical materials for interconnecting the thermocouple to the sensing unit, this problem can be avoided. That's why it is so important to know not only what type of thermocouple is being used, but the correct wire that is also in use. This problem continues back to the reference junction which also may require temperature compensating circuits to maintain calibration. Readings in millivolts are typical, and each corresponds to a specific temperature for a type of junction. The values are usually published in tables by manufacturers. Figure 19-17 shows some types of cases used to enclose thermocouple and RTD units.

The RTD unit (resistance thermal detector) operates in the same manner as a thermistor except that three wires are required for interconnections. Figure 19-18 shows a typical circuit with an RTD unit. The three leads make it possible to balance the circuit since the two wires that are joined at the same point are equal in length.

Type	Conductor materials		Color code	
	Positive (+)	Negative (−)	Positive (+)	Negative (−)
E	Chromel	Constantan	Purple	Red
J	Iron	Constantan	White	Red
K	Chromel	Alumel	Yellow	Red
T	Copper	Constantan	Blue	Red

Figure 19-16 Thermocouple color code.

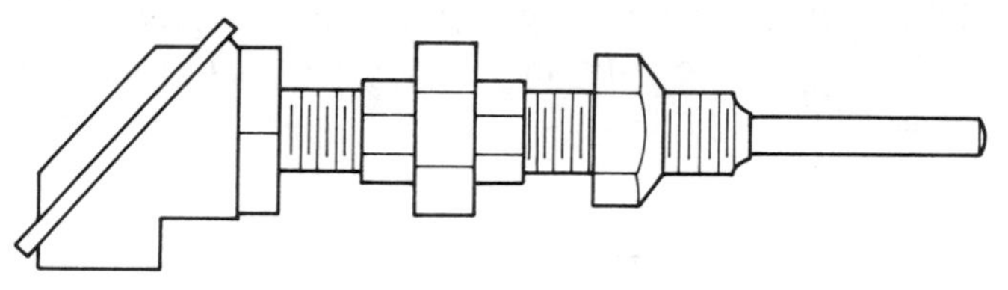

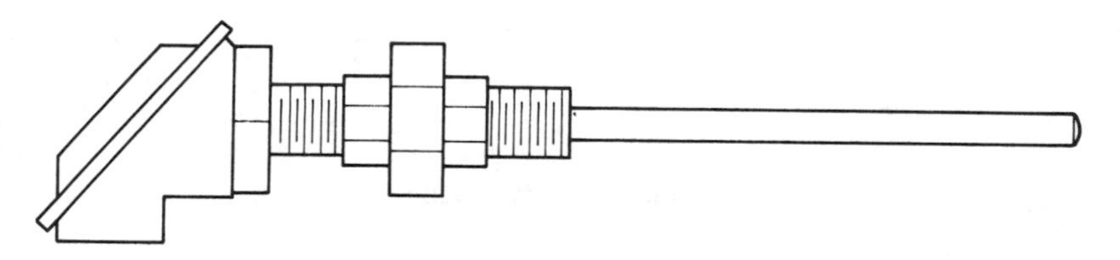

Figure 19-17 Temperature sensor enclosures.

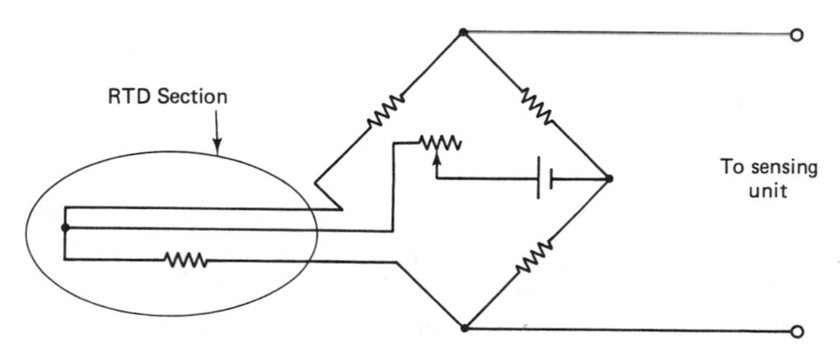

Figure 19-18 RTD circuit.

Radiation Measurements

One additional sensing unit deserves noting. This is the radiation sensor. Two important applications are level sensing and thickness control. For both uses, Figure 19-19 illustrates how radiation units might be placed. In one example a hopper is filled until it blocks a radiation beam. This radiation is at a very low level, and of course, its use is regulated. The receiver unit is calibrated for a specific amount of radiation. If for any reason this beam is blocked, the radiation level will decrease, and the receiver unit will sense the loss and react accordingly. In a slightly different use, a material that can block some portion of the radiation is placed between the transmitter and receiver. Paper or plastic film are good examples. If the thickness would vary by some small degree, the level of radiation will change. This change can be monitored and in turn control some type of corrective measure for a process.

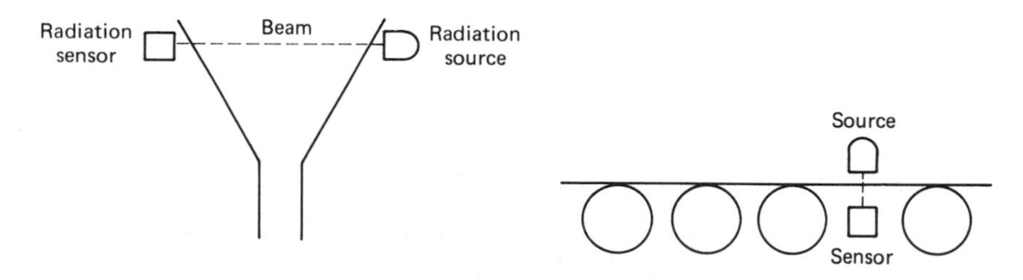

Figure 19-19 Radiation monitors.

CONTROLLERS

The term controllers is used to describe the actual part of the circuit that receives the sensed signal. They may use mechanical, fluid, or electrical devices. Although we might feel that electrical means are the major source of controllers, the pneumatic or fluid types are very popular. Pneumatic controllers can respond with the same actions in close-loop control as electrical units. Each signal coming into the controller is combined, and a result is determined. The result is then used to manage the operation of the system. Signal values from pneumatic controllers are in the 5 to 15 psig range, and electrical signals in the 4 to 20 mA range. In-depth study of controllers is beyond the scope of this chapter, but we hope you can see how extensive the control process can be as well as some of the important elements that are used to sense the inputs to these units.

Two components of the new microprocessor type should be mentioned, however. Computers do not function with analog-type signals. In order to be processed, signals must first be digitized. This is done with an analog to digital convertor (A to D). The operation involves turning the converter ON to sample the input at definite intervals. This enables the input to appear to be in digital form. The use of special circuits that remember the last level of the input can prevent glitches from occurring when the converter cycles on and off. Instead of having spiked waveforms, a stair-step type of signal is maintained. This is the type of signal the microprocessor can use. Microprocessors are not just used for inputs; they also control outputs. The control elements work best with analog-type signals, and for this reason digital to analog convertors are necessary. These circuits take the digital outputs and smooth them before applying them to the final control elements.

SUMMARY

In this chapter we examined the control process in much the same way as we began in the earlier chapters: by looking at the total system and then adding the peripheral component. Details of circuit operation are essential when working at a component level. Your ability to visualize a control system must begin with the block diagram and then progress down to subsystems and finally to the component level. We introduced several theories of operation in simple terms for this reason. Knowing that a component is not completing its function will be sufficient cause to suspect it as the source of the problem. The component can then be replaced or repaired. You must know what proper operation is in order to identify that it is not functioning. We did expand our discussion on thermocouples slightly, since it is very important that proper connections and wiring be used or erroneous outputs will occur. We will next focus on the subject of robotics and how it relates to control systems.

EXERCISES

On a separate sheet of paper, complete a response for each question, statement, or problem listed below.

19.1. List the two major control system classifications.

19.2. What type of interaction exists between the input and output of an open-loop control system?

19.3. Provide an example of an open-loop control system.

19.4. What is a closed-loop system?

19.5. The component that combines the input and feedback signals is called ______.

19.6. List the two forms of feedback.

19.7. Referring to Figure 19-5, of what importance are the actual values of each resistor?

19.8. List the three types of control.

19.9. Describe the basic concept behind proportional control.

19.10. What is the main advantage of the proportional control?

19.11. What does rate control do?

19.12. How does rate control respond to changes?

19.13. What is a transducer?

19.14. List two disadvantages of the linear variable differential transformer.

19.15. List two examples of a mechanical transducer.

19.16. What parameter would you use a Bourdon tube-type instrument to measure?

19.17. List the three most important temperature-sensing elements.

19.18. List two applications for radiation measurements.

19.19. When referring to transducers, what does the term controller stand for?

20

Robotics

OBJECTIVES

Upon successful completion of this chapter, you should be able to

(1) define the term *robot*
(2) discuss functions of the robot
(3) discuss the three major subsystems of a robot
(4) list the various kinds of robots and compare their performance capabilities
(5) discuss the different forms of control used in robotic systems
(6) list the languages available for programming robots
(7) discuss future-generation robots and their systems

INTRODUCTION

Simple control systems are at the heart of many more complex systems. The different types of control circuits that we have studied in the previous chapters are key to understanding the basic principles involved in robotic circuits. Robotics is merely an extension of control circuits to a more specific purpose. A robot may be programmed to perform a single function or multiple functions. Newer robots may even include the ability to make some decisions based on the data that is presented. The extent of the control capabilities of robotic systems will surely expand in the future; this is an area that will have increased importance. Even the term *robot* is not widely accepted. The phrase *automated manufacturing systems* is preferred in many cases. In one chapter, it is not possible to describe properly all of the components or requirements necessary to make or maintain a functioning robot. Entire books are devoted to this. Instead we will highlight the major implications that control circuits have on their use and selected topics that we have already described in the text.

ROBOTIC FUNCTIONS

Robotic units serve many functions. Some of the more common are welding, painting, and parts placement. Most of these units do not appear as the human-form robots that we see in the movies but have either a central base with one extending manipulator arm or are built into a rectangular overhead frame that reaches down to perform its work. Space exploration is one of the best examples of the extent to which robot units can operate by telemetry signals. Undersea exploration with robots allows explorers and scientists to go to depths not possible with human submersibles. Police have been using these units to defuse bombs safely from a distance. Scientists handle radioactive material or explore hazardous environments with the use of robot units. Not all robots have such glamorous functions. Some units perform equally valuable functions as aids to the handicapped—this area will expand as technology progresses. Finally, many robots serve as toys for amusement. Even remote-control models belong in the family of robots. Model airplanes have employed these basic principles for many years. Remote-piloted vehicles (or RPVs) are used for military applications. These are aircraft that can be flown without the need for a pilot in the unit. They are flown as remote control units. This allows high-speed targets for weapons-system testing.

Whatever function they perform, they still contain the basic control procedures, which we have presented, that make it all possible. We do not want to give the impression that all robots are simple machines. The degree of interaction between the various systems can become quite complex. It is this interaction that lets robots perform human-like operations.

ELECTRICAL SYSTEMS

A system that seems to be necessary for all robots is the electrical system. They could not function without it. Portable units rely mostly on DC sources due to their portability. Fixed-frame units in a manufacturing setting may use either AC or DC sources. Alternating current sources are cheaper to bring to the unit, and rectifiers can supply the necessary DC power to aid in providing the precise control that is possible with DC motors. Portable units suffer the same problems as the space vehicles in that conservation of energy practices need to be employed to provide energy for all the equipment in use. A choice must be made either to use low-energy-consuming devices or limit the usage of high-energy items. A delicate balance should be reached. This problem may decrease as newer and better power sources are developed.

Using higher efficiency equipment (that considers the physics of gear ratios and mechanical advantage) can allow the use of smaller devices and thus conserve energy. This is most critical when a drive system must be installed to move a large unit. The type of job you wish the unit to perform will also influence the size of the unit. The robot must be able to lift or provide enough torque to do the job or it will be worthless to us. As the load requirements increase, the size of the unit will also grow.

Using parts that are designed for airplanes can also be a way of reducing equipment size. Size and weight considerations are always a factor in the aircraft industry. However, a change in voltage or frequency requirements may be necessary to accomplish this.

Fixed frame units do not have this problem, since power is readily available, and they are sized to perform at the appropriate level for the loads they are designed to lift. These units also can be oversized as a safety factor without much added expense. Level surfaces are the norm, and this reduces the drive requirements for moving the system. Small motors may not be required, but size reduction is always a consideration in any manufacturing plant. Every square foot or meter of the plant is valuable real estate, and equipment that is small makes it more economical to operate the plant on a dollar/square foot basis.

Solenoids and switches that are used should also be sized according to the electrical load requirements of the system, as well as the physical forces that are applied to them. The smaller robotic systems use microswitches for most control functions. These are ideal, since they are small in size and can handle currents of small magnitude as well as some of the surges that may be experienced. Large units, of course, require higher current and physical limits. It is still not uncommon to find microswitches enclosed in the larger units that provide the protection needed yet the sensitivity that the microswitch can provide.

Solenoids are used to provide movement. The coil is energized and the core element is moved in response to it. The force required must be provided by the solenoid. In addition, the length of stroke must be sufficient. A solenoid capable of moving a 1 in./lb load a distance of one-half inch would not be adequate to move a 2 in./lb load a distance of one-half inch or more. Careful consideration should be given not only to the load but also the distance that a solenoid must move.

MECHANICAL SUBSYSTEMS

Gears, belts, and chain drives are used to improve some of the capabilities of the robotics unit. While these systems can be used as drive means to provide mobility, they may also make the necessary torque conversions needed to move larger loads. Speed changes are another feature we can benefit from by using the same units. Belts are usually found only on the smaller robot systems due to their relatively cheap cost. Chain drives are expensive and are primarily found on the very large robotic units or where gear-ratio advantage outweighs the expense of larger direct-drive motors. Rubber wheels of appropriate size can also function like gears. Direct-gear drives seem to be the most compact systems available. They are not subject to the wear that belt drives experience and also are more reliable.

The freedom of movement that a robot unit has is very important. Try straightening your right arm and turning it without bending your elbow. As you will see, you have a definite range of motion, ROM, possible. This is not quite 360 degrees. We have the same problems with robots. Due to the physical construction of the units, some range of motion problems may be encountered. If a turning or twisting motion is desired, your hand must turn its full limit, and then the process must be repeated to turn an object more than 360 degrees. By selecting different types of joints, we can extend the ROM of the robotic unit. By placing a rotating device with a socket, similar to an electric drill, on the end of the robot arm, we can extend the ROM to an unlimited amount in both directions. In some instances hydraulic systems may be incorporated.

One factor that must be learned early in the study of robotics is that they operate in a three-dimensional plane. We unconsciously move our bodies in three dimensions, but we tend to think in two dimensions when it comes to machines. We must then include three dimensions in our ROM requirements. This can cause added problems when we try to combine all of them at the same time. Wires are much like body parts, and twisting them to extremes will cause them to break. Rotating electrical connections like slip rings might be necessary. This type of connection can cause problems, and we will many times accept the limitations of ROM on our machine and work around them.

Joints also work as levers. The amount of mechanical advantage present is determined by the placement of the fulcrum. The ability to move an object is not only determined by the type of lever but by the direction of the force applied. If we are aided by gravity, the force needed is reduced. If we oppose it, the load on the joints, the motors and other components will be increased. We must therefore be aware of the physical limitations of our machine.

MANIPULATORS

Manipulators are the hands of the robot. These can be two-point, three-point, or multipoint types. Two-point systems are best suited for flat surfaces like sheet steel. This is like using your thumb and index finger. Three point grippers are used for objects that are not flat. They can accommodate almost any shape and are easily used to pick up any object. They utilize the same principle as the old three-legged tables in that three points can find a balance easier than four points. By controlling the speed and the pressure applied to the manipulators, a great deal of force can be applied or delicate functions can be carried out with the use of sensors. The type of manipulator chosen must also interact with the ROM possible from the machine. Manipulators can also be coupled to respond in one-to-one relationships with specific inputs such as those used to handle radioactive materials in a lab. These are valuable in situations where human interaction is required to a great extent but is too hazardous for humans to be present.

The other choices are merely used to match the user needs when a process is done on a repeated basis. Suction cups and electromagnetic plates are available. Specialized tools may be placed on the robot arm if the robot has a very specific function that will not change frequently; spot-welding units are a good example of this type of application.

ROBOT CONTROLS

Nearly every system we have examined so far has a potential use in robotics. The simple on-off or start-stop circuits, the reversing circuits, the speed control of motors, the interfacing circuits, and the special circuits are only a few of the many control applications. Many more exist. Sensors play a very important role, and we will describe just a few more to emphasize some uses not previously mentioned.

When we do not want our robot to strain its arm, we might include a strain gage to interpret the amount of weight or strain that is present. When the amount exceeds the safe limits of our unit, the strain gage will sense this and either turn the system off or alert an operator of the condition. This alert might be a light display or statement on a computer monitor. Another method might be a simple current sensor to determine when a motor exceeds its current rating.

We may wish our robot to have some sort of vision. This might be a response to lights placed in the operating space of the robot. Placing a light-receiving unit in a robot can allow it to sense the presence of a light source. This may be used to position the robot's receiver for maximum light return, or if both transmitter and receiver are placed in the robot, the reflected light might be used to stop the robot's movement (to indicate an object is present).

More advanced techniques might employ the use of ultrasonic or laser range-finding systems. Ultrasonic units use high-frequency radio waves. By transmitting a constant frequency waveform, any change in frequency indicates a change in distance from the transmitter to the object. This is called the *doppler* effect. As long as the object stays in the same position, no frequency change occurs. If it gets closer, the frequency will increase initially and decrease as it moves farther away. Lasers use a similar principle except light is used and it is timed. Applications for these include maintaining proper distances for painting or keeping a welding rod on the work material even as the rod shortens.

Image systems will play an increasing role in future robotic systems. By taking the image and digitizing it and then comparing it to an ideal picture that is stored in the computer memory, corrections can be made. Color analysis of the image may also aid in determining overheated tools and excessive temperature.

Microprocessors are used routinely with robotics systems. They are ideal, since they can be programmed to repeat a process many times. Operations of a repetitive nature must be placed in the program in the appropriate sequence. Again, the human mind compensates at an incredible speed for the actions we perform, and this allows us to pick up a glass off a table or place a tray on a moving conveyor in a restaurant with little or no problem. Consider if you had to write down each step of your actions, one by one, and enable a machine to repeat this process. The position of each component must be described in relation to all other objects. This is done with a three dimensional coordinate system. Now, this can be done in several ways. One method is to move the robot to the positions required and then record them in a computer. This is not sufficient though, since we must also open or close our manipulator prior to picking up an object, close it before we move the arm, and release it at the appropriate place. How fast the process is completed must also be considered. The position of the joints in each axis needs to be recorded also. And any movement while in a position should also be accounted for in the program. This is all possible—and those motions that are repetitive in nature function very well once the program is developed. Next, adding the motion of the conveyor must also be included. As you can see, the process gradually increases in complexity. You would need to apply programming skills along with the knowledge you have of control systems.

Our final comment on control systems again relates to safety. Robotics units, in spite of the degree of sophistication we build into them, will always lack the human element to recognize a dangerous situation—unless we provide the control means or add safety features. It is possible for a control system to fail, and in the case of a robot, this failure can lead to actions that range from a complete halt in operations to wild, erratic actions. The lack of attention at all times by humans may also place them in a position near a robot that would be extremely dangerous. For these reasons nearly all robot units are interlocked to shut down when personnel are too near the machine. This is done by door interlocks, pressure-sensitive mats, limited access by enclosing the system in a cage or a combination of any or all the preceding. Other systems are also in use to provide the same degree of isolation between people and machines.

FUTURE SYSTEMS

Robotics is moving away from the one-unit fixed-program system. New interfacing systems will allow on-line programming and correction of specifications as they are required during the manufacturing process. Dimensions may have to be changed when a problem is discovered. Reprogramming twenty-five or more robot units would result in excessive down time. By using on-line programming methods and interfacing with a central computer, each unit can finish its last product, and then new dimensions can be added to the program by computer interaction between the robotic microprocessor and the mainframe computer. This will be made possible by new protocol, such as the General Motors Manufacturing Automated Protocol System (or MAPS). This system will allow using many brands of robots or controllers with one mainframe computer and will allow the manufacturers an opportunity to buy the best-suited robot or microprocessor for their needs.

Expanding this slightly, if you consider the time it takes to transmit pertinent data to one set of robots in one plant, consider the time involved in providing the data to many different plants in several cities. By using the MAPS system, this data can be sent to each individual machine in every factory accessible by the mainframe computer. This converts to a substantial savings in cost in making new production drawings, mailing expenses, and the number of bad parts produced during the transition time. A simple change in part numbers is easily accommodated by such a system.

Computer Aided Drafting and Manufacturing systems, CAD/CAM, allow the design and testing of the manufacturing process before actually using the tooling process. Designers can use monitors to draw a new part, program the robot, and study the effects of each step. Drawings can be made from the monitor diagram and sent to the supervisor or operator at the machine. If a change is made, a new drawing can be quickly processed and redistributed. Coordination of the entire manufacturing process is feasible with such a system.

Quality control will also influence the use of this system, since it will allow corrections to be made at a faster rate with less waste. Reprogramming of individual machines as the need arises without affecting the entire process can allow retooling times to be reduced.

Accountability of each process is also enhanced. Only the required number of parts needed will be produced. Inventories can be assembled from a variety of sources. Interaction can be done at the management level with data that reflects current production effort. Troubles in specific machines can be isolated. Trends and histories can be compiled. Replacement parts can be ordered on all items, including electrical systems, as they approach their failure rate time or during scheduled preventative maintenance. All of these factors are possible with such a system.

SUMMARY

Control circuits are a very important part of any machine or process. We have taken the common circuits and presented them in what we hoped would be simplified terms for you. Your knowledge of basic circuits should enable you to understand more complex circuits. By giving you a variety of control subjects, we may also have touched on an area of specific interest to you.

EXERCISES

On a separate sheet of paper, complete a response for each question, statement, or problem listed below.

20.1. What is a robot?

20.2. Describe two common functions of the robot.

20.3. With exploration in space, what kind of signals are used to control robot operation?

20.4. What type of switches do smaller robotic systems use?

20.5. List two primary subsystems of the robotic system.

20.6. In what reference plane do most robotic systems operate?

20.7. Relatively, what function of the human body does a robotic-systems manipulator perform?

20.8. What part of the human body is the robot gripper similar to?

20.9. For what purpose are microprocessors used in a robotic system?

20.10. Describe a typical application of a robot's vision system.

20.11. How can a robot recognize a dangerous situation?

20.12. List two processes that may be influenced by future robotic systems.

Appendix A

General Laboratory Safety

INTRODUCTION

Safety is one of the major elements of good laboratory management, but it is perhaps the most neglected. The notion that accidents will always happen to the other person or in their laboratory, and not ours, prevents us from realizing the importance of preventive measures. To be effective in laboratory management, safety must be taken very seriously.

The main objective of a safety program is to make the laboratory or the workplace as safe as possible. The extent to which this objective is achieved depends on how well the causes of accidents are understood. Two major cause categories are listed below.

(1) Human causes: (a) carelessness, (b) ignorance, (c) wrong attitudes, or (d) negligence.

(2) Contributory causes: (a) physical condition of student, (b) unsatisfactory or inappropriate tools, (c) unsatisfactory equipment, (d) dangerous materials, (e) diverted attention, (f) congested workstations, or (g) inappropriate attire.

By means of adequate instruction and motivation, most of these human causes can be controlled. Contributory causes can also be controlled (including environmental conditions).

In this Appendix the important rules, guidelines, and recommendations are presented to help you have a safe learning experience and develop proper employment safety skills. A total program that includes activities of safety education, demonstrations, field trips, safety posters, and so forth will be provided by your instructor. Standards of safe electrical wiring are assumed to have been followed during construction of the school building. This should not prevent regular inspection of each electrical outlet, fuse box or circuit breakers.

TEACHERS' POINT OF VIEW

Teachers are managers of their laboratories and should be in charge of each activity. In this regard, it is recommended that your teacher provide you with fundamental descriptions of first aid and lab evacuation procedures, just in case someone is injured.

THE OVERALL LABORATORY FACILITY

A good shop or laboratory layout and organization plays an important role in the maintenance of a better-than-average safety program. Similarly, a good safety program will be ineffective because of an unsafe shop or laboratory layout. Equipment should be placed in appropriate locations with safety in mind, and consideration should be given to effective installation or repair procedures. Regularly-used equipment, tools, and materials should be stored where they can be easily accessible. All electrical outlets should be connected through a master switch system to a circuit breaker, for emergency shut-down and over-current protection.

Since almost all electronic instruments are sensitive to temperature and humidity, adequate and comprehensive ventilation systems should be adopted.

GENERAL SAFETY RULES AND GUIDELINES

The general safety rules and guidelines recommended for most electronic shops and laboratories should be part of the handout given to you on the first day of class. Specific rules for particular equipment and restricted areas will be explained by your instructor. Some fundamental rules include:

1. Students must comply with all rules and regulations established by the school.
2. Each student must observe all prescribed rules with regard to his or her conduct and behavior in the classroom.
3. Defective equipment or tools should not be used. Tell your teacher if something seems dangerous or does not work properly.
4. No student should operate any equipment in the laboratory before or after school when the teacher is not available for supervision.
5. Due to distractions, two students should *not* be allowed to operate one piece of hazardous equipment at the same time.
6. Students must not be allowed to remove any protective feature from equipment.
7. Students must not use any hazardous or sensitive equipment unless you've demonstrated an ability to operate the equipment properly and your teacher has given permission.
8. Most electrical equipment should be turned OFF and unplugged after use.
9. Hand tools, such as a soldering iron, should be kept clean or sharp and in good working condition.
10. Students should not administer first aid without first getting the instructor's approval.
11. All students should fulfill their housekeeping responsibilities.
12. All accidents should be reported to the teacher, regardless of severity or size.
13. Equipment should be checked by the teacher before and after use.
14. All heat sources must be kept away from computers, data terminals, and other heat-sensitive equipment.
15. No student should attempt to repair any defective equipment without receiving a specific request or instruction from the teacher.
16. Students must not eat, drink or smoke in the laboratory.

These preceding regulations are a partial listing of the safety rules that should be followed during your shop or laboratory activities. Because accidents may occur any time, consideration must be given to the method of handling whatever happens; related first aid or evacuation procedures are a part of post-accident activity.

FIRST AID

A standard first aid package must be available for each laboratory to provide for all minor unforeseen accidents. Serious accidents need to be cared for at the hospital. The use of the first aid facility is the sole responsibility of your teacher. Some rules regarding the use of first aid materials are listed below.

1. The teacher should check the first aid box periodically to replenish used-up items.
2. Ambulance personnel may be called in to give lectures to students regarding the importance of first aid.
3. For extreme emergencies, some administration official should be responsible for transportation of severely injured students to the hospital, if an ambulance is not available.
4. All eye injuries must be treated by a doctor.

HOUSEKEEPING PRACTICES

Good housekeeping reduces the chance for accidents. However, the success of good housekeeping depends on organization and student compliance with established rules and regulations. The physical condition of a laboratory has a strong effect on the amount and quality of learning that may take place. Well-planned and well-managed facilities tend to create good student attitudes, without which effective and creative learning may seem impossible. In fact, uncleaned and disorganized surroundings have a negative effect on learning.

SAFETY WITH ELECTRICITY/ELECTRONICS

Specific rules to follow in order to avoid electrical accidents include the following:

1) Always use equipment with the third-wire-ground system.
2) Use only one hand when probing electrical equipment.
3) Avoid touching any grounded objects while working on electrical equipment.
4) Discharge capacitors before handling them.
5) Do not probe into **Off Limit** components or control panels.

6) Keep the hot tips of soldering irons in a shielded holder.
7) When using an oscilloscope, connect the system under test to the line voltage through an isolation transformer.
8) Use rubber gloves and an apron when handling storage batteries and CMOS devices.
9) In case of fire, turn off the power source first; then attend to the fire. Call the security office or the nearest fire station if the fire is out of control.

Appendix B

Troubleshooting

Troubleshooting, or finding a fault in a circuit, is a skill that takes time and patience to develop. There are two basic concepts that should be understood and used if you are to be successful. First, all short circuit measurements will be zero volts. For our discussion, assume all power source voltages to be 120 VAC. In the circuit illustrated in Figure B-1, the 120-VAC power source is connected to a load consisting of one 120-ohm resistor in series with a switch, SW1. Placing a voltmeter across points 1 to 3 will obtain a reading of 120 volts, which is the power source voltage. At points 1 to 2 it will also read 120 volts. Since we have a complete circuit, the 120-ohm resistor develops the entire source voltage. At points 2 to 3 a meter will read 0 volts. That is because the measurement is across a closed switch, which is a short circuit. This agrees with Ohm's Law and Kirchoff's Law for voltages developed in a circuit.

If we examine the circuit with switch SW1 open, a different set of voltages are obtained with the voltmeters. The circuit has the same components, but the open switch appears as an infinite resistance value to the voltage source. In essence, there is a series circuit with a 120 ohm resistor and a second resistor with a value of infinity. If it were possible to assign a value to infinity, you could theoretically calculate the current of this circuit. If you would like to calculate

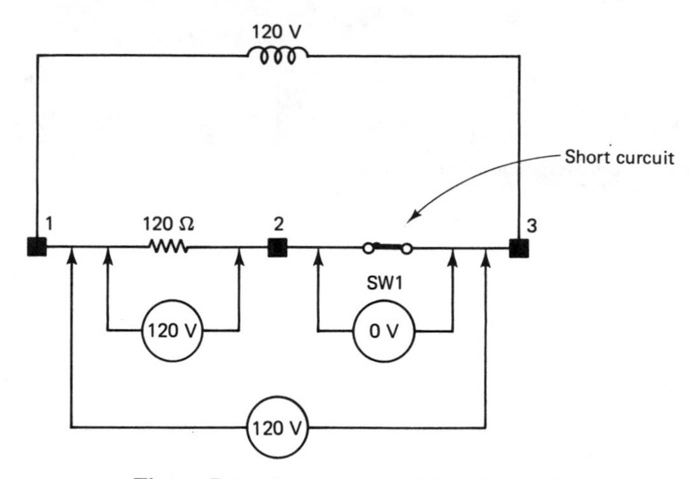

Figure B-1 Short circuit with swl closed.

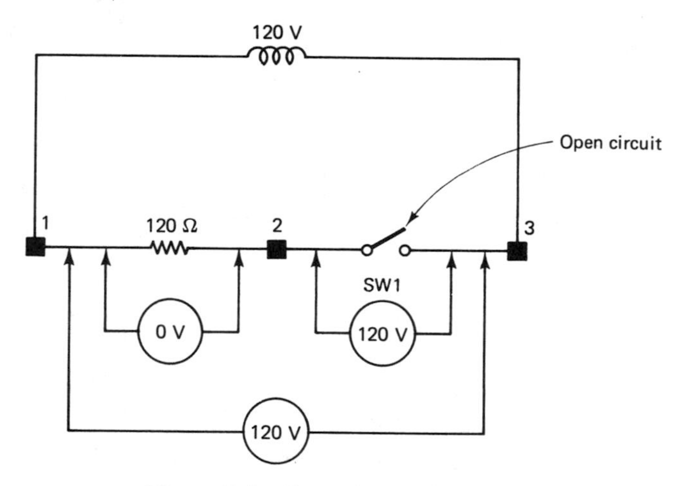

Figure B-2 Open circuit with swl open.

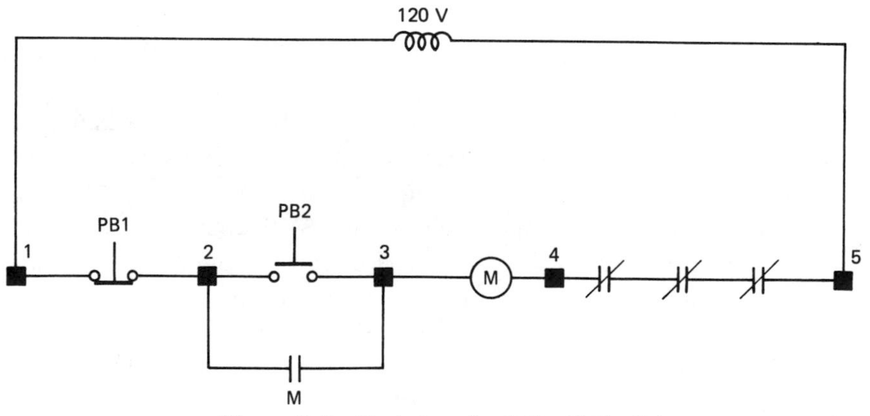

Figure B-3 Start-stop circuit for Table B-1.

current and voltages for this circuit, assign an arbitrary value of 1,000,000 ohms for the open circuit. The 120 ohm resistor would have an approximate value of 0.000119 volts (very near 0 volts) and the open switch (a 1,000,000-ohm resistor) would have a value of 119.999 volts. This for all practical purposes is 120 volts. From this illustration you can perhaps better understand that all open circuit voltages developed across a power source will be equal to the source voltage and all short circuit voltages will be zero volts. Points 1 to 3 therefore will be 120 volts for two reasons. First, it is a reading across the source and secondly because the meter is also reading the open circuit voltage developed across the switch. Without the meter in the circuit, no current will flow and the readings are of no value to us. Points 2 to 3 is the open circuit which develops the 120 volts of the source. Points 1 to 2 is the 120-ohm resistor in series that develops an insignificant value which we can call zero volts. (See Figure B-2.)

If you understand these principles and the need to have a complete circuit path for current to flow, troubleshooting control circuits will become easier for you. Many times it is the meter that will provide the necessary component to make a complete path for current to flow. There are also situations in which no complete path is made, even when the meter is placed in the circuit. For this reason always examine the circuit to determine if a zero-volt reading is due to the lack of a complete circuit and not a short circuit.

To assist your learning of how to troubleshoot a control circuit, a typical start-stop circuit is illustrated in Table B-1 with the voltage readings that should be obtained when it is de-energized and energized. Voltage indications are given along with an explanation for each set of points measured.

TABLE B-1 VOLTAGE INDICATIONS OBTAINED WHILE TROUBLESHOOTING A START-STOP CIRCUIT

Point	Voltage	Indication Circuit De-energized
1–5	120 volts	Normal 1. 120 volts from power source. 2. Open circuit voltage across points 2 and 3. Abnormal 1. 120 volts across any open circuit.
	0 volts	Normal—Only with control power off. Abnormal 1. Loss of control power. 2. Short circuit between points 1 and 5.

TABLE B-1 *(Continued)*

Point	Voltage	Indication
		Circuit De-energized
1–4	120 volts	Normal 1. 120 volts from power source. 2. Open circuit at points 2 and 3. Abnormal 1. Any other open circuit between points 1 and 4.
	0 volts	Normal—Only with control power off. Abnormal 1. Short circuit between points 1 and 4. 2. Normal indication when overloads are open.
1–3	120 volts	Normal 1. 120 volt control power. 2. Open circuit voltage across points 2 and 3. Abnormal—Open circuit at points 1 and 2.
	0 volts	Abnormal 1. Short circuit points 2 and 3. a. Contacts M closed. b. PB2 contacts closed. 2. PB1 contacts are normally closed.
1–2	120 volts	Abnormal—PB1 contacts are open and remaining circuit has been shorted.
	0 volts	Normal 1. PB1 closed, short circuit voltage. 2. No complete path for meter circuit.
		Circuit energized
1–5	120 volts	Normal 1. 120 volts source voltage. 2. 120 volts developed across coil M. Abnormal—No abnormal conditions will exist if the coil is electrically energized, since all remaining points will be short circuits. This assumes that PB1, PB2, and contacts M are functioning as intended and have not remained closed due to possible arcing.

TABLE B-1 *(Continued)*

Point	Voltage	Indication Circuit De-energized
1–4	120 volts	Normal 1. 120 volts source voltage. 2. 120 volts developed across coil M. Abnormal No abnormal conditions will exist if the coil is electrically energized, since all remaining points will be short circuits. This assumes that PB1, PB2, and contacts M are functioning as intended and have not remained closed due to possible arcing.
1–3	0 volts	Normal 1. PB1 is closed—short circuit. 2. Contact M is closed—short circuit. Abnormal No abnormal conditions will exist if the coil is electrically energized, since all remaining points will be short circuits. This assumes that PB1, PB2, and contacts M are functioning as intended and have not remained closed due to possible arcing.
1–2	0 volts	Normal—short circuit voltage across PB1. Abnormal No abnormal conditions will exist if the coil is electrically energized, since all remaining points will be short circuits. This assumes that PB1, PB2, and contacts M are functioning as intended and have not remained closed due to possible arcing.
2–5	120 volts	Normal 1. Control power voltage. 2. 120 volts developed across coil M. 3. PB1—0 volts—short circuit. 4. Contact M—0 volts—short circuit. 5. Overloads—0 volts—short circuit. Abnormal No abnormal conditions will exist if the coil is electrically energized, since all remaining points will be short circuits. This assumes that PB1, PB2, and contacts M are functioning as intended and have not remained closed due to possible arcing.

TABLE B-1 (*Continued*)

Point	Voltage	Indication Circuit De-energized
2–4	120 volts	Normal 1. Control power voltage. 2. 120 volts developed across coil M. 3. PB1—0 volts—short circuit. 4. Contact M—0 volts—short circuit. 5. Overloads—0 volts—short circuit. Abnormal No abnormal conditions will exist if the coil is electrically energized, since all remaining points will be short circuits. This assumes that PB1, PB2, and contacts M are functioning as intended and have not remained closed due to possible arcing.
2–3	0 volts	Normal 1. Contact M—0 volts—short circuit. 2. 120 volts developed across coil M. 3. PB1—0 volts—short circuit. 4. Overloads—0 volts—short circuit. Abnormal No abnormal conditions will exist if the coil is electrically energized, since all remaining points will be short circuits. This assumes that PB1, PB2, and contacts M are functioning as intended and have not remained closed due to possible arcing.
3–5	120 volts	Normal 1. Control power voltage. 2. 120 volts developed across coil M. 3. PB1—0 volts—short circuit. 4. Contact M—0 volts—short circuit. 5. Overloads—0 volts—short circuit. Abnormal No abnormal conditions will exist if the coil is electrically energized, since all remaining points will be short circuits. This assumes that PB1, PB2, and contacts M are functioning as intended and have not remained closed due to possible arcing.

TABLE B-1 *(Continued)*

Point	Voltage	Indication Circuit De-energized
3–4	120 volts	Normal 1. Control power voltage. 2. 120 volts developed across coil M. 3. PB1—0 volts—short circuit. 4. Contact M—0 volts—short circuit. 5. Overloads—0 volts—short circuit. Abnormal No abnormal conditions will exist if the coil is electrically energized, since all remaining points will be short circuits. This assumes that PB1, PB2, and contacts M are functioning as intended and have not remained closed due to possible arcing.
4–5	0 volts	Normal—overloads—short circuit.
	120 volts	Abnormal—overloads tripped or open circuit. *Note:* This condition would appear only if PB2 were held in a closed position manually. The circuit could not remain energized.

There are times when it is helpful to disconnect the power or load wiring from a control device and then test its operation. If there is normal control circuit functioning without the load, then you should consider the possibility that a fault exists either in the load itself or with the wiring from the control device to it.

Ohmmeter checks of relay contacts can also be tested with voltage sources removed and the relay coil circuit energized and then de-energized. You must be cautious with an *open circuit* indication while doing this, since occasionally an oil film may develop on the contacts themselves and a high voltage applied to the contacts will allow them to conduct normally. This problem usually occurs on low-power sources. The ohmmeter has a low voltage source that may not be strong enough to penetrate the oil film and therefore you read an open circuit.

Motor starters should be examined for tripped overloads. Some units have visual indicators of a tripped condition, but most have a manual reset button that must be pushed to reset the overload unit. When the reset button is pushed after a trip condition you will generally hear a clicking type sound as the metal arm of the overload assembly slides across the overload solder wheel (refer to Figure 16-6). If no sound is heard, then the unit may not be tripped. At times you must give the solder time to cool in order for the trip unit to remain closed. Some units incorporate a manual or automatic selection for the overload unit which will reset once the unit cools. A repeated release of the overload unit indicates a problem with the current load applied to the overload assembly. This may be either excessive currents at the motor, incorrect size of overload heaters, or, occasionally, excessive ambient temperature for the selected heater units. The motor-rated current and a heater size chart should be used to locate the problem. Again disconnecting the motor leads and operating the stators will indicate if the problem is the control circuit or the motor, assuming the proper size heaters are in place.

A quick test of motors can be done by measuring between each of the three leads of three-phase motors. If all of the readings are nearly the same, usually something less than ten ohms, and it is not shorted or grounded to the frame, there is a good possibility that the motor wiring is not the problem and the motor is functioning normally. Motor bearings and control wiring could be a possible problem. On new installations, always confirm that the motor wiring connections match the line voltage source applied to it with the nameplate data. Excessive current readings or burned up motors may be the end result of mismatching.

Capacitor start motors can sometimes be checked by removing the load connected to the shaft, applying the normal voltage to the motor, and very cautiously giving the shaft a slight rotation in the proper direction. If the starting capacitor is defective, the motor may start to run normally. A capacitor testing circuit like the one in Chapter 15 may be used to test the starting capacitor or a replacement unit may be installed. The centrifugal switch circuit should also be checked.

Transformers can be checked in a manner similar to motors. Isolate each set of windings and check for an equal resistance reading between each winding. Don't forget the DC resistance is much smaller than the AC, or reactive resistance, when an alternating current is applied to transformers and motors.

Do not let simple items like a burned out light bulb consume a great deal of your time while searching for some exotic solution to a simple problem. Look for a simple solution to circuits that have been functioning normally for some time. Do not go looking for a wiring problem first, since it must have been wired correctly to begin with in order for it to even operate.

Usually the best place to start checking a circuit is the power source. If there is no power, it cannot operate. Fuses are either open or short circuits. A good fuse will read zero ohms out of the circuit and zero volts with a voltmeter placed across it in an energized circuit. A defective fuse will read an open or infinity out of the circuit and possibly a voltage reading across it in an energized circuit. If more than one fuse makes up a circuit and all are defective, it is possible to get no reading across a single fuse. For this reason always confirm that line voltage is present entering the fuse. Turn off the power source before replacing any fuse. If a fuse repeatedly blows, recheck the load amperage for the proper fuse size or remove the load from the fuse circuit and check for a short circuit with an ohmmeter. Once a short circuit has been cleared, then replace the fuse. If you don't check first for a short circuit, you can replace fuses all day long and not solve the problem. Next, look for switches in the OFF or wrong position. And of course look for units not being plugged into an electrical outlet.

If a problem seems to be difficult to solve, have someone else go through the circuit with you, if possible. The expression "Sometimes you can't see the forest for the trees" can be very true after a difficult day or spending a long time searching for a solution to a malfunctioning circuit. Stop and take a short break away from the circuit for a minute and then go back to it. This is helpful sometimes. Above all, knowing when and where a zero volt and source voltage reading should occur in a circuit will save you a considerable amount of time solving a problem. Break large complex diagrams down to one rung of the ladder at a time and check for common problems with this circuit before going from one page to another on a multi-page control diagram. Check for interlocks and relay contacts in other parts of the circuit when they are located in the ladder rung with a malfunction. Sometimes it is possible to manually close relays and starters to check for operation in the energized mode. Do this cautiously, and only when there is no danger of personal injury or property damage, especially with a line voltage connection energized to the device. When possible, check component operations with the power off.

You should also assume the testing device is approved for the voltage rating you are trying to measure. Always select the highest rating and turn the meter range to a more appropriate level if you are not sure of the voltage source in the control circuit. This can only be done if you know the maximum voltage present in the facility you are working. Never assume any voltage range until you have contacted experienced personnel in an unknown work area to find out what voltage sources are present. Only use a high voltage tester on sources greater than 480 volts. If possible, always work with someone—especially with 480 volts or greater. Since this is not always feasible, practice good common sense and safety practices, such as testing your voltage tester on a known voltage source to confirm that your tester is operating appropriately before going to a live circuit, only to find your tester is no good. No problem is worth fixing if it means being injured or losing your life. With good common sense and experience, troubleshooting control circuits can be easy and enjoyable.

Industrial Control Electronics

Index

L

M

N

O

P

R

S

T

U